THE SCIENCE OF TV'S

the BIG BANG THEORY

D0376471

THE UNAUTHORIZED GUIDE

EXPLANATIONS EVEN PENNY WOULD UNDERSTAND

DAVE ZOBEL

ECW

With over-comma'ed gratitude
to
Peter Ward Fay

CONTENTS

FOREWORD
PRESERVING YOUR EPONYMITY

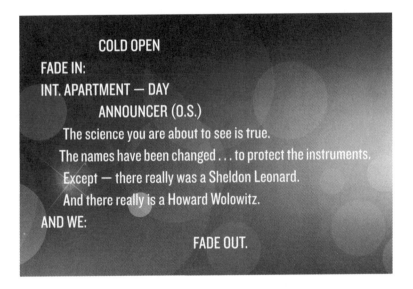

COLD OPEN

FADE IN:

INT. APARTMENT — DAY

        ANNOUNCER (O.S.)

  The science you are about to see is true.

  The names have been changed . . . to protect the instruments.

  Except — there really was a Sheldon Leonard.

  And there really is a Howard Wolowitz.

AND WE:

             FADE OUT.

In 1979, having bought one of the first TRS–80 Model II microcomputers in New York City, I set out to write some commercial software for it. Finding that the only thing it would do reliably was to crash, I was advised to consult a nineteen-year-old genius working at a Radio Shack in Manhattan. His name was Bill Prady, he did indeed turn out to be a genius (with a wicked sense of humor), and within a year he was a vice president and partner at my small computer company: The small Computer Company.

Originally we worked out of my Brooklyn apartment, but once we had grown big enough to move into a proper Manhattan office, Bill showed that he was not only a technical and comedic genius but a resourceful one as well. He draped a pink angora sweater over the back of the receptionist's chair. For years, visitors and customers simply assumed that our receptionist was merely away from her desk, never suspecting that she didn't actually exist.

Comedy proved to be Bill's true love. He started performing standup in the evenings and eventually landed a writing job at Jim Henson Associates. When the company moved to Hollywood in 1990, Bill went with them and began to leave his mark all over the world of television.

One day he asked whether I'd mind if he named a TV character after me. Producer Chuck Lorre had loved Bill's idea for a TV show based on the zany people we both knew from the computer software business, and it seems Bill had always felt *Howard Joel Wolowitz* was the perfect name for a nerd. (My mother would be so proud.)

Chuck and Bill agreed that a roomful of people staring into computer screens might not be the best way to keep an audience entertained. So the characters were changed to be physicists and engineers — which also meant that they could be given real-world scientific problems to struggle with. This book shows how successful that effort has been.

Bill and I remain friends to this day. He even gave me my fifteen milliseconds of fame, in the form of a cameo during the diner scene of "The Re-Entry Minimization" (Season 6, Episode 4), where I can be glimpsed over my namesake's shoulder, seated near a sign that shrugs (appropriately): Sorry — NO CREDIT.

And in case you're wondering, the similarities end with the name. I'm no ladies' man, and dickeys and skinny pants are not for me. But even into my seventies, I'm still making money programming, using the same software that Bill and I developed with others long ago. And I couldn't be prouder to be associated with his homage to the nerds both of us knew and loved.

HOWARD JOEL WOLOWITZ*
*(REALLY)

Newtown, Connecticut
March 2015

INTRODUCTION

WHADDYA MEAN "EXPLANATIONS EVEN PENNY WOULD UNDERSTAND"?!

> [Scene: An apartment building in Pasadena, California, USA.]
> Sheldon's mother: Sheldon, when is your landlord going to fix the elevator?
> Sheldon: I don't know. Lately we've been talking about converting it into a missile silo.
> Leonard: Your son seems to think we need to launch a preemptive strike on Burbank.
> Sheldon: Get *them* before they get *us*.
> — "The Rhinitis Revelation" (Season 5, Episode 6)

What's this? A deadly dance of mutual apocalyptic cease-and-desist, Southern California style, city against city? The left brains of the California Institute of Technology (Pasadena) versus the right brains of Warner Bros. Entertainment (Burbank)? *Why haven't we been warned?*

> **Pasadena** Where *The Big Bang Theory* is set.
> **Burbank** Where the *Big Bang Theory* set is.

In reality, the situation isn't quite that dire. Neither Pasadena nor Burbank has expressed any desire to wipe the other off the map anytime soon. It's only a scene from *The Big Bang Theory*, the situation comedy with the highbrow pedigree.

By turns hilarious and poignant, the show explores the differences — and similarities — between book smarts and people smarts. It features four intellectually gifted social maladroits and one street-savvy ingenue (or, in co-creator Chuck Lorre's facetious turn of phrase, "four wise guys and a sexy dame").[1] And every time the guys haul out their advanced science degrees and start talking shop, viewers know they're in for a buzzword bath.

Except that it's not buzzwords. Nearly every bit of science mentioned on the show is entirely legitimate. It's just that not much of it gets explained. Nor should it. This is CBS, not PBS, after all, and viewers aren't tuning in to be educated, only entertained. "Brilliant people being foolish" is a time-tested formula for comedy, and "foolish people being brilliant" for drama, but it's just not all that funny when brilliant people are brilliant. (And if the thing you most want to watch is foolish people being foolish, well, you hardly need a TV for that.)

Still, it's a rare viewer who doesn't occasionally long for a little more background in whatever it is the characters happen to be jabbering about. Wouldn't you like to know, for instance,

* what Leonard does in the laser lab all day?
* why Sheldon is so fanatical about being "the scientist who confirms string theory" when, according to Leonard (on his very first date with Penny), "you can't prove string theory"?[2]
* where Howard — who, according to the sign on his door, works in an ASTRONAUTICAL ENGINEERING lab in season 1 but a MECHANICAL ENGINEERING lab (and a "RESTICTED" AREA, no less) in season 5 — found the time to become an expert programmer as well?[3]
* when Raj — domestic *divo* by day, stargazer by night — sleeps?

We don't have answers to all of those questions, but we can take a poke at some of them and see what we find. It's not a requirement, of course. You can get a terrific laugh from the nerds on the show without having any background in science at all, but it's nice to know why the nerds in the audience are laughing too.

Okay, "nerds" is unfair — we should probably be saying something like "gifted and highly educated persons." And you can be a scientist without being a nerd. (In a more serious moment, Lorre proposed that the show was "not about geeks or nerds [at all, but] about extraordinary people."[4])

There's an unfortunate myth about scientists, probably fueled by the nerd/geek stereotype. In the popular imagination, scientists are self-assured but introverted. They're walking encyclopedias but ignorant of pop culture. When you ask them how they are, they'll analyze the question in depth before giving a medically accurate response. When you comment on the weather, they'll respond with strange mumblings about global thermodynamics and ripples of causality.

But there are many things scientists know just as little about as the rest of us. Ask a scientist what energy is — or time, or space, or fields, or free will — and watch him squirm. Although the word *science* comes from a root meaning "knowing," science isn't really about *having* knowledge; it's about *searching* for it. (Until the mid-1800s, it wasn't even called science; it was called natural philosophy, which translates roughly to "the pursuit of knowledge about how Everything works.") And it's an endless search. The smartest people know that what they know is nothing compared to what they don't know.

Fortunately, you can understand some surprisingly heady concepts even if you don't think of yourself as having any scientific aptitude whatsoever. Science is all about curiosity, and that same curiosity that drives scientific research runs deep in all of us. Have you never wondered what makes logs float but rocks sink? Why no two snowflakes are alike? Whether bears can ever be taught the rudiments of beekeeping? In the pursuit of knowledge, curiosity is the most important factor.

As for "Explanations Even Penny Would Understand," how often have you said, "Ah yes, it's all clear to me now," when it really wasn't? We've all occasionally faked our way through a discussion that's become convoluted, confusing, or inadequate. It's a way of keeping the conversation moving along and hiding our ignorance, though it often has the opposite effect. (When Sheldon challenges Amy to defend an extraordinary leap of logic, and she responds simply, "Isn't it obvious?" and he counters after an awkward pause, "You're right. My apologies," it's not entirely clear that they aren't both completely faking it.[5])

The admirably unabashed Penny is more apt to say something direct, like, "Okay, sweetie, I know you think you're explaining yourself, but you're really not."[6] She knows that the responsibility for making something understandable lies not with the listener but with the speaker. It's not a question of "dumbing things down," using shorter words, or glossing things over. It's a question of giving your audience the tools they need to help them draw their own mental pictures.

This book is an attempt to give you just a few of those tools. (Where it fails, the fault lies not with Penny or with the reader but with the author.) So come along, and let's delve into what's fun — and funny — about science, engineering, and *The Big Bang Theory*. And who knows? Along the way, maybe we'll learn just a little about how Everything works, starting with the book itself, and the sidebars featured throughout it:

[SCIENCE TO COME] These three little words are there to draw attention to what some people would call nitpicks.

Not many television comedy writers are science experts (big surprise). But so what? Sheldon's whiteboards could be covered with pure kindergarten scribble-scrabble instead of actual equations and diagrams with "a little string theory doodling around the edges," and only a few highly tech-savvy bloggers would grumble.[7] If *The Big Bang Theory* were drowning in the same sort of buzzword technobabble that weighs down reruns of *CSI* and *Star Trek: Voyager*, not many viewers would care.

But it isn't. Laudably, creators Bill Prady and Chuck Lorre wanted their show to be beyond scientific reproach, so they brought in an expert to catch errors, filter out buzzwords, and provide authentic terminology. The startled-looking stranger Howard descends on in the cafeteria with the announcement, "This is my girlfriend, Bernadette,"[8] is David Saltzberg, Ph.D., science consultant to *The Big Bang Theory*. Saltzberg, a real-life professor of particle astrophysics at UCLA, is responsible for most of the show's science references as well as at least one joke (the one about Galileo and the Pope).[9] In a sense, his work is what has made this book possible.

LEONARD
I think you'll find this pretty interesting. I'm attempting to replicate the [science to come].

Those three little bracketed words are how a sitcom writer shrugs. They show up at least once in every script draft, and it's Saltzberg's job to replace them with bona fide dialogue. He clearly enjoys the challenge and especially likes adding highly technical jargon "that you think is Hollywood science but find out not only is it real, it's topical."[10] He's responsible for several mouthfuls of dialogue, including the show's "honorificabilitudinitatibus" (the longest word in Shakespeare): Sheldon's declaration that he once spent a year "examining perturbative amplitudes in $n = 4$ supersymmetric theories, leading to a re-examination of the ultraviolet properties of multi-loop $n = 8$ supergravity using modern twistor theory."[11] This techno-litany sounds almost as over-the-top as his exhaustive recitation of a list of videogame titles,* but it's entirely as legitimate as his hilarious "spoof of the Born-Oppenheimer approximation."[12]

With the greatest of respect, this section points out some of the exceedingly rare moments when the science on the show doesn't quite ring true.

EUREKA! @ CALTECH.EDU
I go for The intensity of Calculation
Leonard, Sheldon, Howard, and Raj may all be fictional (or at best composites), but their employer is very real, and the research going on there (in the real world) eclipses anything on Sheldon's whiteboards.

---

* And outdone only by the title of IRS Form 5213: *Election to Postpone Determination as to Whether the Presumption Applies That an Activity Is Engaged in for Profit.*

Caltech (one word, one capital)** is the California Institute of Technology. Together with the Jet Propulsion Laboratory, which it founded and operates, it's the largest employer in Pasadena. Notwithstanding an undergraduate population of less than 1,000 and a graduate population not much larger, nineteen of the school's former students have gone on to win Nobel prizes, while almost as many faculty members have also been Nobel laureates.

Caltech has undergone a few name changes since its founding in 1891 as Throop University (with a silent H, as in Caltech), but it is not now, nor has it ever been, Cal Tech, Cal-Tech, Cal Poly[tech], or the Southern California Institute of Technology. (Nor, fortunately, Throop Institute of Technology.) It's not a Milpitas auto body shop, a San Antonio software firm, or a calibration company in Quebec. (And you probably didn't need to be told that it's not a construction company in East Timor.) And though both fictional institutions were based on it, it's not CalSci (that's *NUMB3RS*) and it's not Pacific Tech (that's *Real Genius*). It's generally called "the Institute," not "the University" — and it's *never* called "Caltech University," is it, Amy?[13]

From watching *The Big Bang Theory*, you might suppose that the researchers at Caltech spend all their time playing pranks on one another, whining about their love lives, and grumbling at the administration. But that's not the whole picture. They're also solving

---

** Ignore the all-too-common two-words-two-caps misspelling found in the season 4 DVD booklet and elsewhere.

some of the most intricate and complex problems mankind has ever confronted — just a few of which we'll look at in these pages.

ASK AN ICON As any fan of *The Big Bang Theory* will tell you, you don't have to be an expert in science, comedy, or visual entertainment to enjoy its unique blend of all three. Then again, wouldn't it be interesting to know what some world-famous celebrities — folks who *are* recognized experts in science, comedy, and/or visual entertainment — think about the show?

Couldn't hurt to ask. So we did. And several of them answered.

OUT TO LANDS BEYOND Each year, Caltech awards bachelor's or advanced degrees to barely 500 people, not all of whom, surprisingly enough, will go on to become programmers, or professors, or physicists.

Believe it or not, many Caltech graduates choose to spend their lives doing something other than designing interplanetary spacecraft, discovering transuranic elements, disproving centuries-old theorems, or memorizing all the dialogue from the *Star Wars* movies. Included among the Institute's 30,000 alumni and alumnae are opera singers, Oscar winners, science-fiction authors, professional rock-and-rollers, farmers, venture capitalists, and CEOs. The last man to set foot on the Moon is an alumnus, as are a former prime minister of Iceland, an International Motorsports Hall of Famer, and one of the authors of *Shakespeare for Dummies.*[***]

Throughout the book, we'll occasionally take a glance at some of the stereotype-busting ways a Caltech degree can be put to good use.

This section's title harks back to Caltech's alma mater, written in 1919 by Manton M. Barnes ('21) during his junior year. It begins in typically grandiose fashion:

> In Southern California with grace and
> splendor bound
> Where the lofty mountain peaks look out to
> lands beyond ...

Twentieth-century automation and automotion soon clouded the skies and rivers of greater Los Angeles until they rivaled those of Pittsburgh, and the words began to carry an unintended irony. By the time stricter environmental standards were adopted, alternate lyrics had already been circulating for some time, opening with:

> In Southern California with smog and
> sewage bound
> Where the lofty mountain peaks are seldom
> ever found ...

[***] Shakespeare and Caltech are old pals. The third person to win the grand prize on the American edition of TV's *Who Wants to Be a Millionaire?*, Joe Trela (Caltech '97), paused on a question from *Henry VI, Part 2*, to give a shout-out to his Shakespeare professor at Caltech, Dr. Jenijoy LaBelle.

IN WHAT UNIVERSE? "I already, um, have your address," purrs grad student Ramona Nowitzki to Sheldon. But the news doesn't faze him, even when he hears it again from Kathy O'Brien, another Cooper super-groupie.[14] In fact, everybody knows where Sheldon lives. The street address of the apartment building where most of the action of the show takes place is announced in several episodes. It's located just to the northeast of downtown Los Angeles, at 2311 North Los Robles Avenue, Pasadena, California, USA.

There's only one problem: That address doesn't appear on any map. That's intentional. The show's creators deliberately avoided positioning the apartment at any recognizable place. You won't be able to loiter outside in hopes of offering Sheldon a lift to the model train store. Sorry.

Nevertheless, a whole online subculture has grown up around pinpointing the building's location by (playfully) assimilating information gleaned from the show. A pointless exercise, admittedly, but an engaging one.

In this section, we'll take a stab at it ourselves. It'll require a little sleuthing, a little logic, and every "clue" we can lay our hands on: not just the occasional tantalizing glimpse out the window or sidewalk dolly shot, but considerably more.

But don't feel cheated if we don't take it too seriously. After all, this is the same show that once featured a hotel room in Geneva with a commanding view of what appeared to be the Matterhorn — nearly 80 miles away. In what universe would *that* be possible?[15]

1. "The Luminous Fish Effect" (Season 1, Episode 4) — on the vanity card that appears during the final 0:00:01 of screen time.
2. "The Monopolar Expedition" (Season 2, Episode 23) and "The Fuzzy Boots Corollary" (Season 1, Episode 3), respectively.
3. "The Jerusalem Duality" (Season 1, Episode 12), "The Transporter Malfunction" (Season 5, Episode 20), and "The Bus Pants Utilization" (Season 4, Episode 12), respectively.
4. "The Convention Conundrum" (Season 7, Episode 14) — vanity card.
5. "The Zazzy Substitution" (Season 4, Episode 3)
6. "The Hamburger Postulate" (Season 1, Episode 5)
7. "Pilot" (Season 1, Episode 1)
8. "The Gorilla Experiment" (Season 3, Episode 10)
9. "The Cooper-Hofstadter Polarization" (Season 1, Episode 9)
10. Karen Heyman, "Talk nerdy to me," *Science* 320 (2008): 740–741.
11. "The Euclid Alternative" (Season 2, Episode 5)
12. "The Bozeman Reaction" (Season 3, Episode 13) and "Pilot" (Season 1, Episode 1), respectively.
13. "The Tangible Affection Proof" (Season 6, Episode 16)
14. "The Cooper-Nowitzki Theorem" (Season 2, Episode 6)
15. "The Large Hadron Collision" (Season 3, Episode 15)

ONE
THE NAMING OF THINGS

Scientists, right up there with lawyers and loan officers, are widely regarded as inveterate hair-splitters. And they are. They have to be. Science is confusing enough without letting sloppy language make it worse. If you're a scientist, you try to call things precisely what they are:

> "Good news — we've successfully detected the Higgs boson!"
> "Ah, yes: evidence of a key component of the Standard Model of particle physics, don't you know."

If you're a non-scientist, you try to call things precisely what they are, and then typically you provide an alternative nomenclature, starting with the word "or":

> "I saw on the news that they've detected this thingamajig called the Higgs boson . . . or something."
> "Isn't that, like, evidence of a super-important part of the Standard Model of particle physics . . . or whatever?"

Whether using the precise terminology or implying that you may not have it quite right, you're acknowledging the importance of calling things what they are. No one on *The Big Bang Theory* embodies this rigorous adherence to verbal exactitude more than Sheldon does; as Penny points out, he loves correcting anyone who "says 'who' instead of 'whom' or thinks the Moon is a planet."[1] That's what makes his retort about Pluto so uncharacteristically un-scientist-worthy. If a group of people are going to talk meaningfully about planets, they'd better be in agreement about what a planet actually is — especially if they're scientists. It's nobody's fault (certainly not Dr. Tyson's) that when the word *planet* was officially redefined, it stopped applying to Pluto.

What is a planet, anyway? Is it just "a big thing that goes around the Sun"? Unfortunately, depending on your definition of "big," that description applies to potentially millions of objects.

Gaia, What Big Eyes You Have!

Caltech has built and operated some of the most enormous optical telescopes ever constructed. A hundred years ago, the biggest telescope in the world had a primary mirror five feet in diameter. Designed by Caltech cofounder George Ellery Hale and installed atop Mount Wilson, overlooking Pasadena, it's still in use today.

Hale went on to create a telescope with an eight-foot mirror, then one with a mirror twice that size. At 200 inches across, this behemoth (the famous Hale Telescope on Mount Palomar, near San Diego) represents the limits of what can be built from a single chunk of glass.

Nowadays, mirrors can be constructed of arrays of individually movable segments, allowing sizes to continue to increase. The current record-holders, the twin telescopes of Hawaii's Keck Observatory (a Caltech collaboration), have mirrors thirty-three feet in diameter. And when Caltech's next collaboration, the Thirty Meter Telescope, is completed, its mirror will measure nearly 100 feet across. Try hiding a dwarf planet from *that*.

Humanity has been down this slippery slope before. In the very early days of astronomy, what people meant by "planet" really was "a big thing that goes around the Sun." For thousands of years, only five or six were known. (There was some disagreement about whether Earth revolved, which Copernicus resolved.)

Then the telescope was invented, and many additional big things that went around the Sun were discovered and named. By the mid-1800s, the number of objects that had been classified as planets had grown to nearly two dozen. Curiously, all the new additions to the list occupied a single region of the Solar System: a ring around the Sun that's now called the asteroid belt. It contains not a few but millions upon millions of tumbling, rolling fragments of rocky space debris, the building blocks of planets. Half its mass is concentrated in four large fragments, yet the largest of these, Ceres, has barely 1% the mass of Earth's moon.

If you glued all the components of the asteroid belt together, they'd make a ball only about a thousand miles across. That's a pretty small ball: it could comfortably squat on the entire Middle East (not that we would ever wish that) without overhanging the edges. Or if you could somehow spread it like cream cheese, you could just about fill all the oceans of Earth to the brim. That's a lot of cream cheese, but it's not a lot of planet.

To continue labeling each newfound big thing that goes around the Sun a planet would have rendered the term virtually meaningless. So astronomers restored sanity by tightening the meaning of the word so that it only encompassed the eight most massive ones. They reclassified Ceres and its puny associates under the new term *asteroid* ("star-like object").

asteroid belt A band between the orbits of Mars and Jupiter littered with orbiting chunks of rocky material, not one of which deserves to be called a planet.

Sensible, right?

The intervening century and a half has brought us back to nearly the identical situation. Out beyond the orbit of Neptune lie a substantial number of objects, many of them tinier than Earth's moon yet bigger than Ceres — big enough to qualify as planets under the new definition. To keep the word from becoming meaningless again, it was necessary to redefine it in a yet more restrictive sense and to introduce a new term: dwarf planet.

The first of these objects to have been discovered was Pluto. Like Ceres and the other asteroids-née-planets of the 1850s, it has played an important role: its misclassification served to demonstrate the inexactitude of the terminology. Tyson and others lobbied hard for a new category of Solar System objects, arguing that it would be in the best interests of science and the public, and a 2006 vote by the International Astronomical Union made it official.

Unfortunately, in the seven decades since Pluto's discovery, the runt of the Solar System had become an oddly beloved part of humanity's mental furniture. Many people strongly resisted its "demotion," with a good portion of their resentment seemingly grounded in reasons that were more emotional than logical. Awkward company for Sheldon to find himself in.

Mike Brown is a professor of planetary astronomy at Caltech. His 2005 discovery of Eris, which appeared to be larger than Pluto but resembled an asteroid far more than a planet, led the IAU to create the dwarf planet category, whose first member is Pluto.

Brown's memoir is unabashedly entitled *How I Killed Pluto and Why It Had It Coming.*

Q: Why are you astronomers being so mean to little Pluto?
Mike Brown: We're not! Like Sheldon, most of us in the astronomical community *like* Pluto.

Maybe it would be easier for Sheldon if he realized that Pluto's demotion isn't Dr. Tyson's fault, or even the International Astronomical Union's. It's the fault of 1930s astronomers. Pluto was misclassified in the first place because we just didn't know enough about the Solar System.

Now, with so much more known, Pluto can be placed in the correct category. I'm sure it's much happier there.

---

1. "The Hawking Excitation" (Season 5, Episode 21)

TWO

ATTO WAY!

Amy's not the first to be baffled by Americans' refusal to embrace what the rest of the planet has long found to be a remarkably straightforward system. It's not because the math is tricky. The entire system is based on multiplying and dividing by ten and by various powers of ten (100, 1,000, 10,000, and so on).

power A mathematical operation based on the notion of repeatedly multiplying a number by itself. Indicated by a small raised number (the exponent) following a regular-sized number (the base). Examples:

3^1 = a three, all alone = 3
3^2 = two threes multiplied together = $3 \times 3 = 9$
2^3 = three twos multiplied together = $2 \times 2 \times 2 = 8$
10^3 = three tens multiplied together = $10 \times 10 \times 10 = 1,000$

And it's not for lack of exposure. Thanks to a concerted push by industry and lawmakers over the past fifty years, it's now easy to find metric measurements everywhere: on soda cans, scales, speedometers — everywhere but in Americans' heads. Yet the public remains devotedly and inexplicably resistant to the system.

Perhaps it's just an entire nation's way of showing solidarity with the many other countries that have also resisted adopting the metric system (namely, Liberia and Myanmar). Or perhaps this is how the public expresses its disdain for the non-English origins of the system or for the many non-English languages represented in it, mostly in the prefixes it uses to imply multiplication and division by powers of ten.

Some of the exponents in the table below are negative numbers. We know that 10^3 means "three tens multiplied together," but what does 10^{-3} mean? It's hard to imagine how you could multiply "negative three" tens together. And what do you do if the exponent is zero? What do you get when you multiply *no* tens together? Zero? An undefined value? Negative infinity? A month of Sundays?*

Long ago, mathematicians agreed to extend the concept of powers to cover exponents less than one. It turns out that the zero-th power of any non-zero number is 1. This may seem completely arbitrary, but it's not; it continues a mathematical pattern. Since 10^2 is one-tenth of 10^3 (that is, $100 = 1{,}000 \div 10$) and 10^1 is one-tenth of 10^2, it makes sense for 10^0 to be one-tenth of 10^1 — in other words, 1.

Continuing the pattern, 10^{-1} should be one-tenth of 10^0 (or $1/10$), 10^{-2} should be $1/100$, and in general all negative exponents should indicate fractions smaller than 1. (You may have noticed that a negative power is the same as 1 divided by the corresponding positive power; e.g., $10^{-2} = 1 \div 10^2 = 1/100$. Weird, perhaps, but the math supports it.)

Note that all the prefixes ending in -a indicate positive exponents and that all the prefixes ending in -i or -o (except hecto- and kilo-) indicate negative exponents. It's as though an -i or -o ending is the equivalent of the English -th (as in micro- = one millionth). The prefix's first vowel is usually short in English, with the exceptions of micro- (long I sound) and pico- (long E sound). That first vowel also nearly always takes the stress. (The most common exception is kilometer: many people say ki*lo*meter. But you can say *kil*ometer if you want to.) Still, none of these prefixes has any real-world meaning unless it's

---

*   While we're at it, shouldn't zero times zero be more than zero $(0 \times 0 > 0)$? Doesn't it seem as though the absence of nothing ought to be *something*?

Metric Prefixes

(There's no need to memorize or even understand this entire table. Scientists and engineers are just about the only people who ever need to go beyond the first two columns of the first six rows.)

Prefix	Value
deca- or deka-	ten $(10 = 10^1)$
deci-	tenth $(^1/_{10} = 10^{-1})$
hecto-	hundred $(100 = 10^2)$
centi-	hundredth $(^1/_{100} = 10^{-2})$
kilo-	thousand $(1,000 = 10^3 = 1,000^1)$
milli-	thousandth $(^1/_{1,000} = 10^{-3} = 1,000^{-1})$
mega-	million $(10^6 = 1,000^2)$
micro-	millionth $(10^{-6} = 1,000^{-2})$
giga-[*]	billion $(10^9 = 1,000^3)$
nano-	billionth $(10^{-9} = 1,000^{-3})$
tera-	trillion $(10^{12} = 1,000^4)$
pico-	trillionth $(10^{-12} = 1,000^{-4})$
peta-	$10^{15} = 1,000^5$
femto-	$10^{-15} = 1,000^{-5}$
exa-	$10^{18} = 1,000^6$
atto-	$10^{-18} = 1,000^{-6}$
zetta-	$10^{21} = 1,000^7$
zepto-	$10^{-21} = 1,000^{-7}$
yotta-	$10^{24} = 1,000^8$
yocto-	$10^{-24} = 1,000^{-8}$

[*]   Giga is pronounced with two hard Gs (as in "giggle") or with an initial soft G sound (as in "jiggle"). In the *Back to the Future* movies, Christopher Lloyd consistently pronounces "gigawatts" as "jigowatts," which many people consider a flub.

Root	Meaning	Origin
deka	ten	Greek
decem	ten	Latin
hekaton	hundred	Greek
centum	hundred	Latin
khiloi	thousand	Greek
mille	thousand	Latin
megas	great	Greek
mikros	small	Greek
gigas	giant	Greek
nanos	dwarf	Greek
teras	monster	Greek
pico	pinprick**	Spanish
penta	five	Greek
femten	fifteen	Danish
hexa	six	Greek
atten	eighteen	Danish
Same as below, but with a silent P		
z + hepta	seven	Greek
Same as below, but with a silent C		
y + octo	eight	Greek

< THE FRENCH REVOLUTION (1799)

< THE INTERNATIONAL SYSTEM OF UNITS (1960)

< ADOPTED IN 1964

< ADOPTED IN 1991

But it isn't wrong to say it that way — only to *spell* it that way, which is what the script does.

** *Pico* can also mean beak, as in "bird's, pinprick caused by the pointed end of a."

attached to the name of a unit of measure, and therein lies the real power of the system.

One of the more radical aspects of post-revolutionary France (besides post-revolutionary France itself), the metric system was intentionally designed to be simple and elegant: simple because it uses a single unit for each type of quantity (no more inch/foot/yard/mile confusion); elegant because units that are seemingly unrelated (what does length have to do with mass?) are defined in terms of one another. For instance, a cubic centimeter is the volume of a one-centimeter-by-one-centimeter-by-one-centimeter cube (half the volume of your little toe, perhaps). Under specific conditions, a very precise amount of water will fit into one cubic centimeter, and the mass of that amount of water (roughly equal to the mass of a paper clip) was defined to be one gram. The liter (or litre in countries with British-derived spellings) was defined to be the volume taken up by a thousand of those grams of water, so a cubic centimeter is also called a milliliter.

Prefixes and units of measure mix and match freely. From their names, it's easy to see that a centimeter is $1/100$ of a meter and a centiliter is $1/100$ of a liter, just as one cent (which is short for centidollar, or it would be if centidollar weren't completely made up) is $1/100$ of a dollar and one French *centime* is $1/100$ of a franc. Since a meter is a little over three feet and a liter is a little over a quart, a centimeter must be a little over $3/100$ of a foot (about a third of an inch) and a centiliter must be a little over $1/100$ of a quart (about two teaspoons).

In the *Big Bang Theory* sendup that appeared

on *Family Guy,* the "Sheldon" character brags, "I'll have you know that I can bench press over 690 billion nanograms."[1] Impressive — until you convert 690 billion nanograms ($690 \times 10^9 \times 10^{-9}$ grams) to 690 grams, which (at about 454 grams per pound) makes a pound and a half.

These prefixes show up in all sorts of contexts. The measure of sound loudness called the decibel is one tenth of a bel (a far less common but equally valid unit).** Your home's electric meter measures by the kilowatt-hour, which is 1,000 watt-hours, or 1,000 times the energy needed to keep a one-watt flashlight bulb lit for one hour.

Originally, the metric system included prefixes for only the first three powers of ten, using variations on the ancient Greek and Roman words for ten, hundred, and thousand. But since the ancient Greeks and Romans rarely needed to count beyond a few hundred thousand, even when counting ancient Greeks and Romans (it was a smaller world back then), they didn't leave us any suitable root words for higher powers of ten. And that's unfortunate, because following World War II, many governments realized that science and commerce needed standard prefixes for million(th), billion(th), and trillion(th).

Beyond a thousand, it's more convenient to think not in powers of ten but in powers of a thousand; hence the use of commas (or, in some countries, periods or spaces) to break up large numbers into three-digit

---

** The bel is named for Alexander Graham Bell and deliberately misspelled to reduce confusion. (Why, are you confused?)

groups. Our word *million* is an artificial construction that translates loosely as "big thousand," while *billion* is meant to evoke the concept of "twice-big thousand," i.e., "big, *big* thousand."

The conversion from a power of 1,000 to a power of 10 is easy; you just multiply the exponent by three:

$$1,000^1 = 10^3 = 1,000$$
$$1,000^2 = 10^6 = 1,000,000$$
$$1,000^3 = 10^9 = 1,000,000,000$$

etc.

The hard part is agreeing on what to call them.

There Aren't 100 Rs in "Hector"

Some words that look like they should be metric measurements aren't: pentameter (in poetry, a line length of five rhythmic feet); pictogram (a visual representation of something); nanotube (a tube whose length is measured in nanometers); Millicent (someone's name) . . .

The word *micrometer*, when the accent is on the first syllable, means a millionth of a meter, in which case it is more commonly called a micron. With the accent on the second syllable, the same word means a device for measuring small distances (small, but much bigger than microns).

And not every possible combination of prefix and unit is commonly heard. The coastline of Hawaii is roughly a megameter long, and three gigaseconds is roughly a century, but you'll rarely hear anyone use either of those

---

words. Which is a pity, because combinations like that give us a terrifically colorful way to describe the timeline that rolls across the *Big Bang Theory* opening titles. Near the beginning, the label "I Billion Years BC" indicates the point where the scale jumps down by a factor of ten, from "gigayears" to "megacenturies." Each tick mark on the left is worth as much as the entire rest of "BC" combined.

A similar factor-of-ten decrease happens at the next three labels, as the tick marks shift to "megadecades," then "megayears," then "kilocenturies." At 100,000 BC we jump down by a factor of 100, skipping past "kilodecades" straight to "kiloyears" (for which there *is* an accepted word: millennium). We then stay in millennia (with the exception of the peculiar typo at "91,000 9C") until 10,000 BC.

Next, "hectoyears" (centuries) take us to (and past) the inexplicable year "0,"* and at AD 1000 the scale drops to "decayears" (decades). After the year 2000, each tick mark is worth a mere two years — or, in the first five seasons of the show, one year.

---

*   Regardless of the adoption dates of various calendars, your take on world events of the time, or whether you start your centuries in '00 or '01, New Year's Eve in 1 BC is followed by New Year's Day of AD 1. Talking about the year "AD 0" would be as meaningless as talking about "the zero-th year" of an era.

Who Wants to Be a Short Billionaire?

For naming huge and tiny numbers (million/millionth, billion/billionth, trillion/trillionth, and so on), most English-speaking regions of the world use what's called the short scale, in contrast to the long scale used elsewhere. Both systems call the number 1,000,000 (which equals 10^6 or $1,000^2$) a million, but after that the naming diverges. The number 1,000,000,000 (= 10^9 = $1,000^3$), which is a billion in the short scale, is called a thousand million in the long scale. When a long-scale user says *billion*, he means a million million (10^{12} = $1,000^4$) — what short-scale users call a trillion.

In many long-scale countries, another word for thousand million (10^9) is milliard. But a million of those (10^{15}) is a thousand billion, not a billiard; that word is reserved for the family of games in which you place some balls and some money on a table and see which disappears first.

There are good reasons for and against each of the scales, and we won't take sides here. We'll simply use the short scale, since that's the one our publisher uses for reckoning royalty payments.

For the International System of Units, adopted in 1960, the scientists flipped through their dictionaries and came up with a set of prefixes based on six visually arresting words: five Greek, one Spanish. (Spanish! Was there really no way of saying "pinprick" in Greek?)

Within four years, two more powers of a thousand were needed, to bring the prefixes up to a billion billion (and down to a billionth of a billionth). This time, however, the wordsmiths turned away from visual imagery and went back to numbers: two of them from Greek and two from Danish. (Danish! How quickly they gave up on Spanish!) Yes, Danish. And why not? For such a small country, Denmark's contribution to the sciences is remarkable. Anyway, any language that could give the world such a marvelous tongue-twister as *rødgrød med fløde*, which every Dane's child can say effortlessly (and no one else's child can), deserves a place on the tongues of the world's scientists and engineers.

The prefixes peta- and exa- were formed by dropping one consonant from their respective powers of 1,000: the *n* from *penta* ("five") and the *h* from *hexa* ("six"). This was an intentional word game, adopted after someone noticed that by coincidence tera- ("monster"), meaning $1,000^4$, is *tetra* ("four") with a consonant removed.

rødgrød med fløde A fruit porridge ("red groats with cream," but tastier than that sounds). Ask a Dane to say it for you, and stand amazed. Then repeat it back to him by saying, "Logroll my father!" with your tongue hanging out.

The latest additions to the list of prefixes, still with their roots in the Greek counting system, play a combined game of "drop the consonant" and "add a one-letter prefix." And how were those prefixes-on-prefixes chosen? Evidently by starting at the end of the English alphabet (English! Like we earned it!) and working toward the front. So as long as you can recite the alphabet backward while counting in Greek and Danish, you should have no problem with numbers between 10^{-24} and 10^{24}.

The Prescience of the Proto-Hofstadter

Long after the diameters of atomic nuclei had been found to be a few millionths of a millionth of a millimeter, there was still no official word for such a tiny distance. So in 1956, physicist Robert Hofstadter (the eponym of Leonard's family name) came up with one. He coined the term *fermi* (in honor of physicist Enrico Fermi) and defined it as a distance of 10^{-15} meters. Fermi, recently deceased, had done extensive work on nuclear processes, and element number 100 (fermium, symbol Fm) had just been named for him.

With the adoption of the prefix femto- in 1964, the distance formerly known as the fermi became the femtometer, whose abbreviation — by a pleasing coincidence — is fm.

And that's all she wrote. There are no official prefixes to signify 10^{-27} (= $1,000^{-9}$) and 10^{27} (= $1,000^{9}$). But it's only a matter of time, because although economists don't have a need for numbers that extreme (yet), scientists might. After all, 10^{27} cubic yards is roughly half the volume of the Sun, while 10^{-27} pounds is about the weight of a single molecule of hydrogen on the Moon.

Unfortunately, the next English letter is X, with

all of the pronunciation ambiguities that that entails. Worse, the Greek for nine is *ennea*, and dropping its one and only consonant sound doesn't leave much more than a donkey's hee-haw (and all that that entails).

Evidently it's time to abandon the "drop a consonant" scheme, which is why the prefixes xenia- and xenio- have been proposed. (These are unrelated to the Greek *xenos*, meaning "stranger," as in xenophobia, the fear of Danish children.) However, other candidates are also up for consideration: xenna-/xenno-, xenta-/xento-, and even the brazenly English-centric nina-/ninto- (with a long I sound).

Alternatively, the caretakers of the International System of Units could take the suggestion of a group of students at the University of California, Davis. Reverting to the descriptive-names theme of mega-, giga-, and tera-, they proposed that 10^{27} simply be given the evocative prefix hella-.

While all of that is being sorted out, if you're an American (or you know one) looking to help further the spread of metric know-how in the States, why not reprogram the GPS in your car to speak in kilometers instead of miles? (Of course you would want to download the British voice pack for that — which, let's face it, probably has a much more soothing cadence than the American one anyway.) You could also learn to give your height in centimeters, your weight in kilos, and your favorite drink recipes in milliliters.

Work on it. Keep it up. You'll be sounding like a citizen of the world in no time.

Things Don't Just Add Up — They *Multiply* Up

As Sheldon laments to Amy, his misreading of square centimeters as square meters has given him a result that is "off by a factor of 10,000." But if there are 100 centimeters in a meter, why wasn't he off by a factor of only 100?

The answer is that a square meter contains not 100 square centimeters but 10,000. Imagine taking a square sheet of paper one meter on each side and cutting it into 100 strips, each one centimeter wide by one meter long. Now imagine cutting each of those strips into 100 one-centimeter-by-one-centimeter squares. Your one square meter of paper will yield 10,000 little square centimeters of paper.

This effect works even more dramatically in three dimensions, and it explains why a thimble can hold only a tiny fraction of the water that a drinking glass can hold.

Imagine you have a hatbox in the shape of a cube. Anything you do to make it shorter (or narrower or squatter) will cause its volume to decrease correspondingly. If you could squeeze the box down to one-tenth its original height without changing its length or width, so that it resembled a pizza box, you could stack ten copies of it (10^1) in the same amount of space as the hatbox takes up.

Squeezing the pizza box down to one-tenth its width without changing its length would give you something like a box of spaghetti, and you could fit ten of them into the space of one pizza box, or 100 of them (10^2) into the space of one hatbox.

And suppose you squeezed the spaghetti box down to

one-tenth its length. Now you'd have a miniature hatbox — cube-shaped just like the original, but only one-tenth its size in each dimension: one-tenth as long, one-tenth as wide, and one-tenth as high. And you could crowd not ten but a *thousand* (10^3) copies of it into that darn hatbox.

EUREKA! @ CALTECH.EDU
Let's Get Small
On the eve of the 1960s, physicist Richard Feynman delivered an address at Caltech entitled "Plenty of Room at the Bottom." In it, he included a challenge to physicists and engineers that quickly became world famous.

Feynman was speculating on the enormous untapped possibilities of machinery designed to operate at submicroscopic scales — what's now known as nanotechnology. The "bottom" in the title of his lecture refers to distances with very large negative exponents (meaning extremely short lengths): below millimeters, below micrometers, below nanometers, down to femtometers — the scale of atoms.

As for "plenty of room," that's no exaggeration. In principle, Feynman observed, the smallest things we can build could be designed to build other things even smaller than themselves. He envisioned a day when microdevices made of individual molecules could be assembled, atom by atom, by the world's tiniest robot hands, in the world's tiniest factory.

You Can't Get There from Here

If we wish to visit the apartment building in which Sheldon, Leonard, and Penny all live, we can start by noting that it's owned by "the 2311 North Los Robles Corporation."[2] By an astonishing coincidence, the address of the building is also given as 2311 North Los Robles Avenue.

Los Robles Avenue is a wide oak-lined street cutting through central Pasadena. Like most other north-south streets in the city, it's numbered starting from Colorado Boulevard (the east-west artery that features in the Jan and Dean song "The Little Old Lady from Pasadena"). The hundreds digit changes about every eighth of a mile, so 2311 North Los Robles should be almost three miles north of Colorado Boulevard. (How "the gas station across the street" from the apartment could be named "the Colorado Boulevard Chevron" remains a mystery.[3] That would have to be one long gas station.)

Two and a half miles north of Colorado Boulevard, at Pasadena's northern border, there are only about three streets that don't continue into Altadena. Los Robles is one of them. A strip mall called Hen's Teeth Square, straddling the city line at 2061 North Los Robles Avenue, is the farthest north you can go.

When Sheldon, through an act of willful vandalism, briefly alters the building number to "311," he shifts the action to a real address.[4] However, there isn't a real building at that real address — just a freeway overpass. The only building-like structure anywhere nearby is a curbside utility box too short for even Leonard to fit into.

We do have the right Pasadena, right? There's a city by that name outside Baltimore, another one near St. Petersburg, and a much larger one in east Texas, not far from Sheldon's Galveston home. But not one of them has a Colorado Boulevard or a Lucky Baldwin's[5] or a Cheesecake Factory or — of course — a Caltech.

Clearly, more investigation is needed. We may need to revisit the problem a little later — assuming we can find our way back to it.

---

1. "Business Guy" (Season 8, Episode 9)
2. "The Financial Permeability" (Season 2, Episode 14)
3. "The Recombination Hypothesis" (Season 5, Episode 13)
4. "The Desperation Emanation" (Season 4, Episode 5)
5. "The Justice League Recombination" (Season 4, Episode 11) and "The Irish Pub Formulation" (Season 4, Episode 6), respectively.

THREE
CAN YOU HEAR ME NOW?

Sheldon: [Leonard and Penny] had very little in common, except for carnal activity. That's why I acquired these noise-canceling headphones.

— "The Plimpton Stimulation" (Season 3, Episode 21)

Poor Sheldon: he'd be lost without that pair of noise-canceling headphones, especially if Leonard and Penny were having a sleepover. Or Leonard and Stephanie. Or Leonard and Leslie. Or Leonard and Joyce Kim.

As for Raj, he also has a pair. (Despite what some people may say.) He has to wear them whenever Leonard and Priya are having a sleepover.

But this raises some important questions: How do noise-canceling headphones work? What happens to the noise when it gets canceled? Does it just dissipate like smoke? Does it bounce back in the direction it came from? If a tree falls in the woods and it lands on a pair of noise-canceling headphones, does it make a sound? And, most importantly, how on earth did people survive during the thousands of millennia between the invention of roommates and the invention of noise-canceling headphones? (Or, since there appears to be a common thread here, perhaps the question is: "In a world in which noise-canceling headphones don't exist, how would Leonard's roommates ever stop throwing up?")

To understand the noise-canceling process, we have to recognize that sound is a pressure wave, consisting of a high-pressure region followed by a low-pressure region, which is followed by another high-pressure region, which is followed by another low-pressure region, and so on. If you could see a sound wave traveling through the air, it would look something like a windblown field of grain, with amber waves of high and low density chasing each other across the fruited plain.

Typically, a sound is caused by the repeated back-and-forth motion of a vibrating object. As the object moves outward a little, it gives a physical shove to anything surrounding it — molecules of air, for example. Those molecules move outward a bit and bump into their neighbors, transferring the shove to them. Those neighbors bump into *their* neighbors, and so on, and a wave of high pressure moves outward.

Meanwhile, as the vibrating object moves back inward, it creates a small amount of suction, drawing nearby air molecules back toward itself. The inward motion of those molecules creates suction on their neighbors, and so on, and although the molecules themselves are being drawn inward, the wave of low pressure moves outward. Then the vibrating object begins moving outward again and the cycle repeats, with air molecules sloshing back and forth as waves of alternating high and low pressure pass through them.

Doing the Wave

A wave is a disturbance, not a physical object. As objects disturb objects near them, which then disturb other objects near *them*, the wave propagates through space.

For example, the circular pattern of ripples caused by dropping a rock into a pond is actually a self-reproducing chain of local disturbances. Water molecules disturbed by the rock disturb their neighbors, which disturb their neighbors, and so on. A disturbance is always transferred to all neighboring points — not just to those in a line away from the initial disturbance — but half of those neighbors have already been disturbed, and the overlapping patterns of disturbance cancel one another out in all directions but one. The result is that a new wave, indistinguishable from the old one, moves outward from the initial disturbance.

The faster the object vibrates, the shorter the distance between successive high-pressure (or low-pressure) regions. That's because the speed of the waves depends mostly on the density of the material they're passing through, rather than on the strength or speed of the vibration. The pitch of the sound is a measure of how rapidly the waves alternate between high and low pressure, while the perceived volume is a measure of the overall pressure difference. If there's only a small pressure difference between the high-pressure and low-pressure zones, the sound doesn't carry much energy and is perceived as soft; if the difference is very large (very high pressure alternating with very low pressure), the sound carries more energy and is perceived as loud.

Sound doesn't travel at all through a vacuum. Without a medium (such as the intervening air) to conduct them, the pressure waves would have no way to reach your ear, and you would hear nothing. Deep space isn't a perfect vacuum, but the few specks of matter drifting through it — mostly lonely hydrogen atoms — are so widely separated that shoving or pulling on any one of them has almost no effect on any of its neighbors. That's why in space, no one can hear you scream. (Or in Leonard's case, plead.)

She Got the Wave to Move Me
The energy carried by a water wave isn't measured by its height alone. If it were, then any ripple in a third-floor water glass would be more powerful than a tsunami in a ground-floor swimming pool. The most powerful waves are the ones with the biggest height difference between crests and troughs: the highest highs, followed by the lowest lows.

This is true of all waves, even those whose intensity isn't measured by physical height. The crests and troughs of a sound wave aren't up-and-down displacements but regions of high and low pressure, with molecules being alternately pushed forward and sucked backward. The bigger the pressure difference between the regions, the more powerful the wave and the louder the sound.

And light waves aren't a displacement of any physical thing at all. They're changes in the intensity of an electromagnetic field, and they point out sideways as the light travels forward. The higher the crests and the lower the troughs, the more powerful the wave and the brighter the light.

The loudest sound your ears can tolerate without damage is millions of millions times more powerful than the softest sound they can pick up. Between those extremes, the alternating pressure changes they can detect all lie within a certain range of frequencies. If the pressure shifts from high to low and back again twenty times a second or so, it comes across as a very low-pitched hum. If it shifts 20,000 times a second, it's perceived as a high-pitched whine. (Leonard again.) Sounds changing more slowly or more quickly than that just don't register on the ear, unless the ear belongs to an elephant or a bat or any of various other creatures.

As we get older, we lose our ability to detect the lowest volumes and the very highest and lowest frequencies, which is why crafty teenagers often select especially high-pitched cell phone ringtones inaudible to adults. Teachers consider this an exceptionally polite gesture, as it gives them a much less disruptive way to be ignored during class.

Haven't You Heard?

The simplest way to test the range of a person's hearing is to play tones of different frequencies and loudnesses and ask whether he can hear them. But when verbal communication is not an option, such as with infants or non-humans, the test must be done by other means.

In 1957, Dr. John C. Lilly (Caltech '38) was training a dolphin to whistle for tidbits. Each time the dolphin whistled — at any frequency — Lilly rewarded it. At one point the dolphin fell silent, so Lilly stopped rewarding it. After a short time it began whistling again, and the rewards recommenced.

On reviewing a recording of the session, Lilly realized that during the silent portion of the concert program, the dolphin hadn't ceased whistling at all. With its ability to produce (and hear) frequencies more than seven times above the highest frequency humans can hear, it had simply been whistling in that hypersonic range. Once a few ultra-high-pitched whistles had gone unrewarded, it chose a lower frequency and never raised it again after that.

Was the dolphin using the tidbit reward system as a way of testing the range of human hearing? Had the subject become the experimenter?

It's easy to see this as a case of turnabout. We humans often rely on experimentation as a way of finding things out, and we know that dolphins are smart creatures, so we might decide that the animal was "curious" about human hearing, "wanted" to "learn" something about it, and "decided" to conduct a "test." But anthropomorphizing isn't an accepted part of the scientific method (see "And *That's* How It's

Done"), and neither is jumping to conclusions. Had the dolphin actively chosen to test Lilly's hearing, or was it just fooling around? Without the ability to read minds, even the most adept dolphin whisperer couldn't know what was going on underneath that blowhole.

In fact, a dolphin smart enough to be a scientist probably wouldn't conduct the experiment in this way, since the results didn't tell it whether its highest whistles were inaudible to Lilly — only that for some reason they weren't being rewarded. This is like a parent who will occasionally buy you ice cream when you ask politely but never (though you're no less audible) when you scream.

Without insight into the dolphin's motivations, all we can really conclude is that it learned (whether it set out to do so or not) that whistling above a certain frequency gets you no reward.

Still, that's pretty useful information . . . for a cetacean.

Since what you're hearing is not the motion of the object itself but a pattern of alternating air pressure on your eardrums, there needs to be an unbroken path from the object to your ears. Anything along this path that reduces the pressure difference reaching your ear will make the sound you hear quieter.

Covering your ears with something rigid that air molecules bounce off, such as a drinking glass, prevents some sound waves from getting through. The longer the time between the arrival of high- and low-pressure regions, the less effectively the object resists them, so low frequencies get through better than high ones.

Covering your ears with something that physically slows down the sloshing of the air molecules, such as a pillow, allows each incoming high-pressure zone to catch up to the low-pressure zone ahead of it. The longer the physical distance between the two, the less completely they overlap each other, so again, low frequencies get through better than high ones.

Loud + Loud = Silent?

Waves can add together. Where a crest meets a crest or a trough meets a trough, the result is a double-strength wave. Where a crest meets a trough, the result is neither a crest nor a trough but an in-between state.

If two waves overlap in such a way that their crests always coincide and their troughs always coincide, such as can happen when two radios placed side by side are tuned to the same station, they add up to produce a double-strength wave. The water rises very high and sinks very low, or the sound pressure is very high and then is very low, or the electromagnetic field is very strong in one direction and then is very strong in the opposite direction. We get flooding, or a loud sound, or a bright light.

But if two waves overlap in just such a way that the crests of one always coincide with the troughs of the other, as happens inside noise-canceling headphones, they add up to produce nothing. The waves cancel each other out, and it's as if there were no wave at all: what one wave giveth, the other taketh away.

Noise is any sound that's undesirable. Noise-canceling headphones are designed to suppress sounds coming toward your ears from outside the headphones themselves (the assumption being that all such sounds are undesirable and therefore qualify as noise). They can't suppress unpleasant sounds that are a part of the audio signal coming from your MP3 player, or that are coming from inside your head.

Noise-canceling headphones come in both passive and active designs. Passive models generally have a hard plastic cover (for reflecting sound waves) over a foam baffle (for muffling the waves that get through).

Active designs are similar to the passive type but are also able to create a sound wave of their own, one especially designed to reduce the pressure of incoming high-pressure zones and increase the pressure of low-pressure ones. Each ear cup contains a small outward-facing microphone that detects sounds coming from outside. Electronic circuitry models how those sounds will be altered by the passive elements of the headphones and then sends the opposite signal to the speaker in that cup. Wherever the first wave has an increase in pressure, the other has a decrease, and vice versa. At every moment the two cancel each other out. (Imagine trying to fill a bathtub whose tap and drain are controlled by the same handle.) The incoming noise, plus its artificially generated opposite, add together to produce silence.

Canceling a sound by adding it to its opposite sound results in no sound, but that's different from not having any sound in the first place. Energy is used up in the process. It's like a perfectly matched tug-of-war: although the rope never moves at all, both teams expend plenty of energy. Every time an external sound tries to move the air-pressure "rope" in one direction, the noise-canceling circuitry moves it in the opposite direction, just enough to cancel the noise's effect. When the microphone detects rising pressure that would push inward on the eardrum, it pulls the speaker outward, lowering the pressure; when it detects falling pressure that would pull the eardrum outward, it pushes the speaker inward, raising the pressure. At every moment, the incoming noise is canceled by the opposing action of the speaker.

That being said, one possible solution to the problem of the noise-emitting roommate is to make an equal noise in the opposite direction. When your roommate gasps, you gasp back. When your roommate moans, you moan back. When your roommate sobs, you sob back. Or you could just stick your fingers in your ears and make gagging noises.

EUREKA! @ CALTECH.EDU
Innovation with (or without) Invention
Though life without his precious noise-canceling headphones would be unbearable for Sheldon, in many ways he's no different from most of us. How long would you last without your smartphone? Your TV remote? Your two-way wrist radio?

Yet to the so-called bottom of the pyramid, the four billion people living on less than three dollars a day, the imported consumer electronics that are so indispensable to us remain an unattainable luxury. For them, the problems of daily living are less "How can I rave about my favorite song in only 140 characters?" and more "How can I afford to run my farm except by selling it?" or "How can I lose fewer elderly relatives to heatstroke this summer?"

A suite of courses created by Caltech professor of mechanical engineering Ken Pickar, with input from the non-profit network Engineers for a Sustainable World, focuses on technological solutions to the problems of developing countries. Effective solutions, Pickar finds, often come from inexpensive mash-ups of existing technologies rather than from the invention of slick new gizmos. For example, some of his students combined a miniature turbine and dynamo into a pocket-sized cell phone charger designed to be hung outside the window of a commuter bus and powered by the breeze. (In the developing world, cell phones are cheap but electricity is expensive.)

There's more involved here than just clever ideas. The world's coolest gadget is worthless if it never leaves the drawing board, so Pickar requires his students to create business plans to address manufacturability, affordability, route to market, and return on investment. For their final exam, the students present their products to an audience of real-world manufacturers, business leaders, and venture capitalists. The result could be the next Intelligent Mobility International, a non-profit spun off by a group of Pickar's students who had developed a method for converting two ordinary mountain bikes into a rugged wheelchair. (In many developing countries bicycles are plentiful, inexpensive, and durable, while medical equipment is not.)

So the next time you hear someone complaining that there's nothing new under the sun, tell him about the bus festooned with whirling phone chargers and the wheelchair with the *awesome* tires.

[SCIENCE TO COME]
Zero to Sixty in 2.74 Seconds

One of Galileo's many contributions to science was to observe that near the surface of the Earth, a falling object accelerates at a constant rate. If the object is initially at rest, it falls four feet in the first half-second (as you can easily confirm by dropping something from that height), twelve feet in the next half-second, twenty feet in the half-second after that, twenty-eight in the half-second after that, and so forth. After one second it will have fallen sixteen feet, after two it will have fallen sixty-four feet, and with every passing second its vertical speed increases by slightly over twenty miles per hour.

He also noticed that this acceleration is independent of the object's mass. Assuming equal air resistance, two objects released simultaneously from the same height will hit the ground at the same time, regardless of their weights. The heavier object doesn't fall any faster than the lighter one does, though you might think it would. (Caltech undergraduates commemorate Galileo's accomplishment at midnight every Halloween by dropping frozen pumpkins of various sizes off the highest building on campus, to fall and shatter in unison nine stories below.)

Adding horizontal velocity doesn't change this effect. If one of the objects is thrown sideways and the other is simply released, both will still reach the ground at the same time. During each second of flight, their vertical speeds increase by the same amount.

Adding vertical velocity doesn't change the effect, either — only the result. If you throw an object upward or downward, its vertical speed still changes by a specific amount each second. The only difference is that, depending on whether you threw up or down, the object will take more or less time to reach the ground.

In what for him is a terribly reckless moment, Leonard pries open the elevator doors outside his apartment and drops a (very recently emptied) liquor bottle down the elevator shaft.[1] Timing its fall by the smashing sound it makes about 2.1 seconds later, he does some quick mental math and announces approvingly, "Thirty feet." Dividing this by the apparent height of the floors in the building (although appearances can be deceiving; see "Says You!") suggests that the bottle has struck something at about second-floor level, perhaps the top of the elevator car.

Except that his mental math isn't very good. In the first 2.1 seconds of free fall, an object — any object — covers a vertical distance of not thirty but *seventy* feet. Whatever his bottle has hit, it's evidently about two stories below ground level.

Sheldon performs a similar experiment but obtains a very different result when he tosses a whiteboard out his living-room window. A mere 1.2 seconds later it crashes onto the street below, causing cars to swerve — which seems a trifle peculiar, given that the window faces the rear of the building. That flight time corresponds to a distance of at most 23 feet (but more likely 10 or less, owing to the whiteboard's not-insubstantial air resistance). This confirms the result obtained by Penny when she hurls an iPod out the same window and it clatters to the ground barely half a second later. Half a second is enough time for gravity to pull a stationary object down a mere four feet. The downward motion of her throw no doubt helps to speed the iPod toward its rendezvous with the pavement, but even taking that into account, it would appear that ground level is somewhere near the middle of the third floor. (On the plus side, this revelation takes the teeth out of Sheldon's threat to Penny: "If you use my toothbrush, I'll jump out that window."[2])

A time-tested movie method for estimating the height of a cliff or the depth of a pit is to drop a pebble into the darkness and listen for it to hit bottom. But if you do that on *The Big Bang Theory,* all bets are off.

---

1. "The Lunar Excitation" (Season 3, Episode 23)
2. "The Einstein Approximation" (Season 3, Episode 14), "The Tangerine Factor" (Season 1, Episode 17), and "The Dumpling Paradox" (Season 1, Episode 7), respectively.

FOUR
I AM THE VERY MODEL OF A MODERN MODEL ORGANISM

Penny: I need some guinea pigs.

Sheldon: Okay, there's a lab animal supply company in Reseda you could try, but if your research is going to have human applications, may I suggest white mice instead?

— "The Grasshopper Experiment" (Season 1, Episode 8)

Poor Penny. She's learning to mix drinks and is only looking for a few people who'll let her practice on them. Sheldon, missing the point as usual, assumes she's looking for a model organism, of which guinea pigs and white mice have historically been crowd favorites.

What gives a species the honor of being known as a model organism? Essentially, it needs to have been the subject of many studies where it acted as a stand-in for one or more other organisms. Amy's smoking monkeys are examples (standing in for smoking humans, presumably), as are Bernadette's small rodents deliberately infected with flesh-eating bacteria (standing in for anything made of tasty flesh).[1]

> **model organism** Any of several organisms that are widely used for performing living studies. The results can then be extrapolated to other organisms.

Another widely used model organism is *Danio rerio,* the zebrafish. This common tropical fish, about an inch long, is gold colored with five dark blue stripes running along each of its sides from head to tail.

But while a zebra is a dark-colored animal with light stripes, a zebrafish is a light-colored animal with dark stripes. (Honestly, except for the stripes, a zebrafish looks about as much like a zebra as a horsefly looks like a horse.) And during its early developmental stages, before its stripes are in, the zebrafish is nearly transparent — one of its main advantages as a model organism. For instance, it's remarkably easy to observe the heart beating.

Even more remarkable is that the heart of an embryonic zebrafish, like the heart of an embryonic human, is an elastic tube with no valves. How does a valveless pump pump? A muscle wrapped around the heart periodically squeezes it. Surprisingly, the squeezing action isn't applied *along* the tube but *across* it. The simple act of rhythmically contracting in and out — squeeze/release/squeeze/release — keeps the blood flowing in one direction.

This is due to the Liebau phenomenon: if one section of a fluid-filled tube is comparatively less rigid than its neighboring sections, then squeezing it periodically, anywhere other than at its center, will induce the fluid to circulate. That's because although the compressed fluid is forced out toward both of the surrounding rigid sections, it experiences greater resistance from whichever of the two is closer, causing a net flow toward the other.

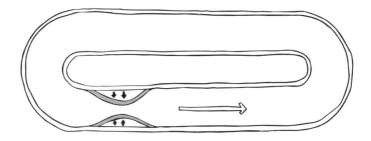

A Liebau tube. Squeezing the less rigid section of the tube off-center sets up a flow.

The zebrafish isn't the only model organism that isn't a mammal. Other familiar faces around the lab include *Drosophila melanogaster* (the ubiquitous fruit fly) and *Caenorhabditis elegans* (the transparent micro-worm with the mega-name).

Experimentation on live non-human subjects (and particularly on adorable live non-human subjects — sometimes referred to as charismatic macrofauna) remains a necessary part of the scientific process. (Sheldon is correct in saying, "There's a *lot* of harm in trying something new. That's why we test out drugs and cosmetics on bunny rabbits."[2]) Yet it strikes some as cruel and unjust, and there are those who would have all testing done either only on humans or not at all. Both of these extremes would be irresponsible. Human experimentation is unavoidable, but it will always be closely scrutinized and heavily restricted.

Whether subjects are being fed sugar pills or cross-bred with gorillas (as in Sheldon's apocryphal Stalin story and Penny's movie, *Serial Ape-ist 2*), substantial issues of ethics and personal rights are involved. Every investigation carries at least some risk of causing physical or psychological damage, in ways that can't always be predicted or controlled.[3]

Of the many acts of unspeakable cruelty littering the history of science, not all have been inadvertent. Some of the Holocaust's worst atrocities took the form of simple but deadly human experimentation, such as when prisoners were placed in ice water to induce fatal brain hemorrhages and thus prove the importance of fur-lined collars for aviators. Professional ethics were deemed less urgent than obtaining knowledge that would undoubtedly save lives. The ends were felt to justify the means.

In 1963, Yale psychologist Stanley Milgram delved deeper into what historian Hannah Arendt, analyzing the behavior of the petty functionaries who made up the bulk of the Third Reich, had termed "the banality of evil."[4] He set up a meta-experiment involving human subjects, invited ordinary Americans to act as laboratory assistants, and found he could easily coax them into committing outright torture and then rationalizing it away.

Subjects can make irrational decisions — but so

can trained investigators. In 1971, Stanford professor Philip Zimbardo set up a mock prison as part of a psychology experiment. He recruited college students to take on the roles of guards and prisoners. To his surprise, bizarre and vicious behaviors quickly surfaced, many evidently motivated by little more than bloodlust and an urge to oppress others. Zimbardo himself became frighteningly seduced by the allure and power of his own role as prison supervisor. After only six days of the intended fourteen, finding that the participants were continuing to mistreat one another and realizing that he could no longer guarantee their safety, he called off the experiment.*

---

* One of Zimbardo's most chilling observations was that although participation was strictly voluntary at all times, not one test subject quit on his own. Every "prisoner," even those being subjected to substantial psychological and physical trauma by the "guards," opted to stay.

EUREKA! @ CALTECH.EDU
A New Wave Extracting Energy

At Caltech's Center for Bioinspired Engineering, mechanical engineer John Meier was researching ways to use the Liebau phenomenon to create artificial microscopic and biomedical pumps having no valves and only a single moving part, but he couldn't help thinking bigger. Could the phenomenon be used as the heart of an alternative energy-capture system?

An appendix to Meier's 2011 Ph.D. thesis describes his design for a novel beachfront device for extracting energy from the ocean. Water circulating through a generator would be kept in motion by nothing more than a periodic "squeeze" from the breaking waves. Simple, perhaps, but as Meier came to realize, simple is not the same as obvious.

Amy's proposal to test the flow of information among her friends by spreading a false rumor seems innocuous, even amusing. But because she can't possibly predict or control the behavior and reactions of others, there's a risk of serious physical or psychological damage. And when Sheldon declares human experimentation to be "one of the few forms of interaction with people that I don't find repellent,"[5] it's amusing in light of his avowed misanthropy.

But on sober reflection, it's no joking matter.

Out to Lands Beyond

"I discovered poetry at Caltech, can you imagine that?" Frank Capra (Caltech '18) directed many of the world's best-known films, including *It's a Wonderful Life* and *Lost Horizon*. In 1915, Sicilian-born Capra entered Throop College of Technology (as Caltech was then known) to study chemical engineering. Though he had the highest grades in the freshman class and graduated in only three years, he found himself in no rush to enter the industry; instead, he wandered through a succession of part-time jobs and gradually fell into film direction.

In the late 1920s, when the technical complexities of the new "talkies" were leaving many directors struggling, Capra's engineering background gave him an important edge. It also informed his 1950s series of educational science films, including *Hemo the Magnificent* (about the circulatory system) and *Our Mr. Sun* (an introduction to solar physics).

These days, it's common to see the director's name promoted above a film's title, but Capra's was the first.

1. "The Agreement Dissection" (Season 4, Episode 21) and "The Psychic Vortex" (Season 3, Episode 12), respectively.
2. "The Rhinitis Revelation" (Season 5, Episode 6)
3. "The Junior Professor Solution" (Season 8, Episode 2) and "The Gorilla Dissolution" (Season 7, Episode 23), respectively.
4. Hannah Arendt, *Eichmann in Jerusalem: A Report on the Banality of Evil* (New York: Viking Press, 1963).
5. "The Herb Garden Germination" (Season 4, Episode 20)

FIVE
HEXAGON WITH THE WIND

Sheldon [to himself]: Unit cell contains two carbon atoms.
Interior angle of a hexagon is 120 degrees.
Howard [to Leonard]: Have you tried rebooting him?
Leonard: No, I think it's a firmware problem.
— "The Einstein Approximation" (Season 3, Episode 14)

Why is Sheldon so frantically drawing and redrawing arrays of geometric shapes? Has he found a new calling in geometry? Does he find pleasure in sketching out his roommate's love life? Is he thinking of giving up on theoretical physics and becoming an architect or (*shudder*) an engineer?

The fields of science and mathematics are much more intimately entwined than many people realize; in fact, they depend on each other for corroboration and inspiration. Many of the revelations of theoretical physics began as mathematical insights. The principles of relativity were worked out not by sending a twin on a rocket voyage or driving a car at the speed of light with the headlights on, but through a clever thought process involving only basic arithmetic. Pencil and paper pointed the way to concepts like antimatter and Heisenberg's famous uncertainty principle long before laboratory tests confirmed them (see "It's a Fuzzy Old World"). The five-decade effort to prove that the Higgs boson exists stemmed from a mathematical analysis that said it should (see "The Naming of Things").

Sheldon's attempt to visualize the flow of electrons through a graphene sheet requires him to revisit the angles and demons of geometry, and there's no shame in that. You're probably aware that angles can be measured in degrees (symbolized by a little raised circle) and that there are:

* 360 degrees in a full circle (as in "a 360-degree view");
* 180 degrees in a half-circle ("he did a complete one-eighty and ended up facing in the opposite direction");
* 90 degrees in one corner of a square or a rectangle.

Some other angles commonly encountered are:

* 120 degrees: any corner of a hexagon (such as a cell in a honeycomb);
* 60 degrees: any corner of an equal-sided triangle;

* 45 degrees: either of the two pointy corners of a square that's been folded along its diagonal;
* 30 degrees: the angle between two adjacent hour markings on a clock face;
* 6 degrees: the angle between two adjacent minute markings on a clock face.

The sweep second hand on a clock changes its angle by one degree every sixth of a second. The minute hand, moving sixty times slower than that, covers one degree every ten seconds. The hour hand moves one degree in a full two minutes. But that's still twice as fast as the Sun, Moon, and stars, which all take four minutes to crawl one degree westward (or rather, to appear to sit still in the vault of the heavens while the Earth rotates one degree eastward on its axis under them). The Moon lags behind the stars a bit, because during those four minutes it creeps about 3% of a degree eastward along its orbit. The Sun also backtracks: during those four minutes it moves $1/360$ of a degree eastward on its annual tour of the constellations of the zodiac.

A satellite or spacecraft in low Earth orbit rises and sets every hour and a half, traveling one degree around the Earth every fifteen seconds or so. And a satellite in geosynchronous orbit doesn't rise or set at all, since it moves as fast against the background stars as the background stars seem to move against the horizon: one degree every four minutes.

geosynchronous Completing exactly one west-to-east orbit of the Earth per day. As viewed from the ground, a satellite in geosynchronous orbit performs a little dance in which it slides north and south against the sky, wiggles slightly east and west, and returns to the same spot daily. Geosynchronous orbit requires a very high altitude, averaging roughly 22,000 miles (a tenth of the way to the Moon, and nearly 100 times higher than the International Space Station). **geostationary** In a circular geosynchronous orbit over the Equator. An object in geostationary orbit appears to hover perpetually over a single spot on the ground, with scarcely any wiggling dance at all.

Each degree of arc (abbreviated °) is divided into sixty minutes (abbreviated '), and each minute equals sixty seconds (abbreviated "). Used as measurements of arc, these have virtually nothing to do with the familiar minutes and seconds with which we measure time. In both cases, the words *minute* and *second* have something to do with division by sixty, but the things that are being divided (time on the one hand, angles on the other) are unrelated to each other.

For instance, in one day (24 × 60 = 1,440 minutes of time) the Earth rotates through 360 degrees (360 × 60 = 21,600 minutes of arc). In two minutes of time, the Moon moves one minute of arc against the background stars.

But so what? It's not like *minute* and *second* are any

more ambiguous than the word *degree* itself. According to *The Big Bang Theory*, there's fine meteor shower viewing from 34.48 degrees north, 118.31 degrees west (as well as great locations for *Star Trek* shoots), Sheldon sets his thermostat to seventy-two degrees, Raj's parents give him the third degree, Amy mentions having suffered second-degree chemical burns on her face in high school, and of course Howard has only a master's degree.[1] Then there's the notion of six degrees of separation, a healthy degree of skepticism, degree 'n' green grass of home — the list goes on.

° ′ ″ The symbol used for minutes of arc (′), as well as for feet, is intended to look like a small raised Roman numeral I. The symbol for seconds (″), and for inches, is supposed to be a small raised Roman numeral II. As for the little raised circle that represents degrees (°), it's tempting to see it as a little raised circle that represents all other circles, but in fact it started life as a little raised zero. Seriously.

The naked eye can't make out angles narrower than about a single minute of arc. That's roughly as much of your (360-degree) horizon as you could block with your thumb . . . if your arm were 100 yards long. Even a whole degree isn't a lot: it's about as much of your horizon as you can block with one pinky finger held up at arm's length. To be able to block out your entire horizon, you would have to have 180 pinky fingers on each hand. (Then again, if you were a person with 360 pinky fingers, or 100-yard-long arms, your horizon would probably be blocked out anyway — by hordes of angry torch-bearing villagers.)

How many pinky fingers would you need in order to block out the entire Sun (or Moon)? Two? Six? Perhaps surprisingly, one pinky finger alone is more than enough. As viewed from Earth, the Sun and Moon are each only about half a degree wide, or half a pinky-finger's width. (Don't believe it? Try it.)

Angles are as vital to chemists as they are to astronomers and watchmakers. When atoms join up to form molecules, they rarely lie in straight lines and flat sheets. The electromagnetic forces among their positively charged nuclei and negatively charged electrons move them into complex positions. An example is the towering molecular model next to the closet in Leonard and Sheldon's living room, representing a fragment of the very famous (and very non-flat) DNA double helix. In fact, DNA not only twists like a warped ladder but also coils back on itself in complex ways.[*] Even good old water — an oxygen atom in a hydrogen sandwich — isn't arranged in a straight line but in a wide V shape that happens to make almost exactly the same angle as a clock's hands at 10:08.[**]

---

[*] This particular DNA model was constructed from two model kits joined end to end. A single cell's worth of DNA would require almost half a billion additional kits and would extend eight times around the Earth.

[**] This is approximately the "happy face" time seen in nearly all advertisements for clocks and watches. No doubt the clock is so happy because it likes teaching you about geometry and molecules. (Cf. Sheldon's sarcastic "Is it the geometry that makes it fun?"[2])

One historical instance in which math pointed the way for science is captured in a sketch glimpsed on both Leonard's fridge and Dr. Gablehauser's whiteboard (although Gablehauser's has an error). In the 1960s, it was noticed that when a hexagonal layout is used to diagram the family of particles called mesons, patterns emerge. The search for an explanation for those patterns led to the discovery of the particles known as quarks.[3]

As for why there are 360 degrees in a circle (rather than, say, ten or fifty or a million), we have Nature to thank for the inspiration: the arc of the sky divides itself into degrees for us in a very visual way. The pattern of stars visible just at sunrise or sunset changes by about one degree per day. That's because the plodding Earth moves about $1/360$ of the way along its almost-circular orbit around the Sun each day. This may not matter much to you, but it was of great importance to your ancestors. Long before they had agreed on which body plods around which, they noted how many days it took the pattern of stars seen at sunset to repeat, and they called that a year. This method was more accurate than tracking climate-based indicators like seasons and flooding.

And since viewing the stars at sunset was really all your ancestors had by way of evening entertainment (and who's to say that what we have is so much more edifying?), they were well aware that a year is a bit more than 360 days long. But it's close to 360, and 360 is particularly convenient to work with. For one thing, it happens to be evenly divisible by every whole number between one and ten (with one exception, which you can work out for yourself). No other number under 500 can make that claim. In fact, 360 is divisible by a whole raft of numbers. That's especially useful when chopping up a circle into two parts (or three, or twelve, or ninety), which happens more often than you might think.

Other than that, there's nothing magical about 360 or about degrees. You could use any system for measuring angles, and it wouldn't change their properties. For instance, the unit called the gon (or gradian) is very much like the degree, except that there are 400 gon in a circle instead of only 360. That makes a gon about 11% narrower than a degree and means that very few angles come out to round numbers in both systems: the 120-degree corners of Sheldon's hexagons are each $133^1/_3$ gon, for example. Gon are used mainly in surveying and by the military, and it's easy to see why. We humans are partial to square corners and to nice round numbers like 100, and what do you know: every corner of a square has exactly 100 gon. That was an intentional decision, brought to you by the same clever folks who devised the metric system (see "Atto Way!").

A third system of measuring angles is called radians. A radian is a pretty big angle (nearly sixty degrees), and there are only about six and a quarter of them in an entire circle. What use is that?

Imagine a rolling wheel. Every time it rotates through an angle of one radian, it moves itself forward a distance equaling its own radius (hence the name). Radians are a way of chopping up the circumference of a circle simply by using another part of the circle, rather than by hauling in an arbitrary number, as we have to do with degrees (360) and gon (400). That's what makes them a favorite of scientists, mathematicians, and those geometry-crazy wheel-rollers Sheldon loves to mock: engineers. When he isn't going geometry-crazy himself, that is.

EUREKA! @ CALTECH.EDU

Form Ever Follows Function

When you're dealing with the large and complex biomolecules called proteins, it's all about the angles. Proteins are the building blocks of the cell: they give it structure, force molecules to react with one another by pulling them into alignment, and relay signals. They're made of amino acid chains that refuse to lie flat and instead fold and curve and warp themselves into fantastical shapes, like the most convoluted twisted-pipe-cleaner critter a kindergartener ever fashioned. And that's a good thing, because it's mostly the physical structure of each specific protein that gives it its abilities.

In recent years, scientists have modified existing amino acid sequences and created never-before-seen proteins for performing specific functions. Azurite, a brand-new, long-lasting, vibrant blue fluorescent protein, was invented by tweaking some of the amino acids in the green fluorescent protein called, helpfully, "green fluorescent protein" (GFP). Direct control over the sequence allowed properties like physical stability and high-temperature tolerance to be engineered directly into the structure.

Whether a protein will behave as desired depends on how it folds itself, which depends on the electromagnetic forces between its amino acids. These are difficult to calculate, and the architects of novel proteins are always looking for shortcuts, but at present, computational protein design remains part science, part art. Combining sequence fragments copied out of libraries of existing components turns out to be a more fruitful approach than just randomly shuffling amino acids around or — worse — starting from scratch. Protein chains are typically hundreds of amino acids long, and with nearly two dozen candidates to choose from for each link in the chain, that's a lot of possibilities to wade through.

Steve Mayo, one of the minds behind Azurite, is the chair of Caltech's division of biology and biological engineering and a master at protein folding. He's developing procedures to automate the process of architecting designer proteins and reduce the amount of luck required.

Once a candidate recipe has been found, it's easy to whip up a batch of it for testing: you simply create a DNA chain that codes for the desired sequence of amino acids, splice it into an *E. coli* bacterium, and let Nature's assembly line start mass-producing the protein. The design is the hard part.

ASK AN ICON: Harold Rosen

Engineering legend Harold Rosen (Caltech M.S. '48, Ph.D. '51) is considered the father of the geosynchronous satellite. Fifty years ago, his Syncom satellites made possible the ocean-spanning telephone and television we now take for granted.

In his eclectic career, Rosen has also invented a radically new automobile engine, developed unmanned aircraft capable of supplanting communications satellites, and pioneered alternative fossil-fuel extraction techniques. He even led a team designing a lunar lander.

Proof indeed that engineers aren't just glorified construction workers.

Q: Sheldon dismisses engineers as "semi-skilled laborers [who] execute the vision of those who think and dream," calling them the "Oompa-Loompas of science."[4] In your experience, is this a common sentiment among theoretical physicists?

Harold Rosen: I think Sheldon's completely alone in that opinion. Most of the theoretical physicists I've encountered — and I consider myself a scientist as well as an engineer — have had tremendous respect for the engineering profession, and particularly for the most creative engineers.

Personally, I've known engineers whose originality equaled that of any scientist: people like Bill Pickering, the long-time head of JPL; John Pierce, who designed the transmitter amplifier used on most spacecraft; Robert Noyce, inventor of the microchip; Joe Sutter, father of the

747; and Frank Whittle, who invented the turbojet engine.

The space program relies much more on engineering than on theoretical physics. The whole semiconductor industry was created by engineers. Those are truly important achievements, and the vision to create them didn't come from theoretical physicists.

1. "The Adhesive Duck Deficiency" (Season 3, Episode 8), "The Bakersfield Expedition" (Season 6, Episode 13), "The Justice League Recombination" (Season 4, Episode 11), and "The Date Night Variable" (Season 6, Episode 1), respectively.
2. "The Peanut Reaction" (Season 1, Episode 16)
3. "The Jerusalem Duality" (Season 1, Episode 12) and "The Luminous Fish Effect" (Season 1, Episode 4), respectively.
4. "The Jerusalem Duality" (Season 1, Episode 12)

SIX
THE GRAVITY SITUATION

> Leonard: That's Penny's ex-boyfriend.
>
> Sheldon: What do you suppose he's doing here? Besides disrupting the local gravity field.
>
> Leonard: If he were any bigger, he'd have moons orbiting him.
>
> — "The Middle Earth Paradigm" (Season I, Episode 6)

It's awfully brave of Sheldon and Leonard to face up to Kurt, a hulking Neanderthal* poetically described by the script for the pilot episode as "frat-boy-turned-stockbroker." He's taller than either of them, more massive than both of them combined, and ripped.

They're right about his gravitational prowess, too. Without getting hung up on what "the local gravity field" is, we can say what it does: it pulls. Every part

---

* And that's being kind. By calling Kurt *Homo habilis* ("tool-using man"), Leonard places him two million years earlier in the evolutionary family album. (Despite the implication in the theme song, tools strongly predate *Homo neanderthalensis*.)

of every object in the room (or anywhere else in the Universe) exerts its own gravitational pull, which reaches out into the space around it via something called a "field." (And what's a field? Let's just say that it's a "thing" that carries the influence of other "things." Not very satisfying, is it?) The more mass in the object, the stronger the pull: doubling the amount of material in the same volume of space would double the pull. Kurt exerts a Kurt-sized pull, and though it may not be enough to pull down a wall hanging or pull up the carpet, it can certainly be measured by a sensitive detector. (He could increase it by increasing his mass — say, by filling his pockets with change. Unfortunately, that costume he's wearing doesn't seem to have pockets. It barely has costume.)

On the down side (so to speak), the strength of his far-reaching gravitational influence drops off quickly with increasing distance. However weak his gravitational pull is at a distance of, say, one mile, at two miles it's only $1/4$ of that strength, and at three miles away it's only $1/9$. This has nothing to do with his mass or what he's made of; it would be true whether he were a supermassive black hole or a two-bit piece of metal. It happens because gravity obeys an inverse-square (or $1/r^2$) law.

Friendly Neighborhood Mispunctuated Spider-Man
Middle-earth is the name of the fictional setting of J.R.R. Tolkien's *The Hobbit*. In the novel, it's hyphenated; in the title of this episode, it isn't.

To the dismay of Tolkien purists, no explanation (or apology) for this unilaterally reformed orthography has ever been made. We can only assume that the same people were responsible for naming "The Bat Jar Conjecture" (Season I, Episode I3), which, in keeping with "Batphone," "Batcave," "Batcomputer," etc., ought to have been spelled "Batjar."

At press time, neither Moby-Dick nor Winnie-the-Pooh could be reached for comment.

To visualize the geometry behind the inverse-square laws, you can use an empty picture frame (just the frame, no backing), a bright lamp or flashlight, a dark room with a blank wall, and a ruler or tape measure.

Choose a measurement on the ruler: ten inches or so works well. Call this length L.

Prop up the picture frame parallel to the wall at a distance L from it. Place the lamp the same distance (L) beyond the frame, allowing it to cast the frame's shadow directly onto the wall. You'll find that the shadow measures twice as high and twice as wide as the frame itself. This is true no matter what distance L you choose.

Now place the frame where the lamp is and move the lamp another L away from the wall (so that the frame is now $2 \times L$ from the wall and the lamp is $3 \times L$ from the wall). The shadow is now three times as high and three times as wide as the frame itself.

Again, place the frame where the lamp is ($3 \times L$) and move the lamp an additional L away from the wall

($4 \times L$). The shadow is now four times as high and four times as wide as the frame itself.

Since the distance between the frame and the lamp is always L, the amount of light passing through the frame never changes. And regardless of the distance chosen for L, whenever the wall is r times as far from the lamp as the frame is, the shadow is r times as tall and r times as wide as the actual frame. In other words, at r times the lamp-to-frame distance, a constant amount of light has spread out to cover an area the size of $r \times r$ (or r^2) frames. That spreading "dilutes" the light by a factor of $1/r^2$.

$99^{75}/_{100}\%$ There

When the frame is much closer to the wall than to the lamp, its shadow is just barely larger than life-sized. If the Sun were a bright pinprick instead of a disk, making daytime shadows sharp-edged instead of fuzzy, at high noon the shadow of the International Space Station (over 110 yards long) would hardly be larger than the ISS itself. Even if the ISS could fly all the way to the Moon, it would have covered only $1/_{400}$ of the distance to the Sun, and its shadow on the Earth would grow barely a foot longer.

Inverse-square laws describe the behavior of light, sound, exhaust from a tailpipe, the spreading cloud of Kurt's cologne, and other things that fan out in two directions as they move along a third. They're what drives the laws of perspective, making a building look half as tall (and half as wide) when viewed from twice as far away and making the Moon appear to be the same size as the Sun (400 times wider but 400 times farther away). (See "KISS and Tell.")

Knowing Kurt's mass, we can estimate his contribution to the local gravitational field, but we'll run into some difficulties. It's not interference from his animal magnetism. It's not a problem of precision, or the fact that gravity is the weakest of the fundamental forces in Nature. (Kurt can easily override gravity's pull. He could use a big magnet to lift a trash can lid, or he could simply turn the chemical energy in his muscles into motion and take a jump, preferably a running one.) The problem is that Kurt, like most objects in our world, is not perfectly homogeneous and not perfectly spherical. From far enough away, any object looks like a featureless dot, and far away from Kurt, his gravitational influence smooths out until it's essentially indistinguishable from that of a sphere of uniform density. (And now perhaps you get Leonard's "spherical chickens in a vacuum" joke . . . sort of.[1] Hilarious . . . sort of.) But up close, it's apparent that different parts of Kurt have different densities, and in fact he's quite lumpy in places. (Penny seems to have noticed this.) For that reason, his local gravitational field is not spherically symmetrical: it has blobs and lobes here and there.

The effects of gravity don't travel outward instantaneously, as Newton assumed they would, but only at the speed of light: roughly 700 million miles an hour. Seven hundred million miles is about the distance between Earth and Jupiter when they're on opposite sides of the Sun. So an hour after Kurt does anything, the gravitational effects of that action will have reached and passed Jupiter, causing (he would doubtless like to imagine) some impressed Jovian eyebrow-raising (or whatever it is they raise out there).

As he lumbers around the room, the only change in his pattern of gravitational influence will be that it

It's Not *All* about r^2

Inverse-square laws are a feature of life in our three-dimensional Universe, but not every example of something flowing from a source obeys them. The flow of ants scattering from an anthill, for instance, decreases in proportion to r, not to r^2, because the ants are restricted to two choices of direction rather than three (they can choose to go left or right and forward or backward, but they don't launch themselves into the air or burrow back into the ground). If you were to set up croquet wickets (or small arches) on the ground at various distances from the anthill, you could count how many ants per second crossed through each one. Assuming the ants are fanning out evenly, a wicket ten yards from the anthill will see $1/2$ (not $1/4$) as much foot traffic per second as one five yards away.

A flow that's restricted to moving along only one dimension doesn't decrease at all. Water flowing through a pipe can't fan out, r or no r, so no matter where along the pipe you look, you'll still see the same amount of water moving past in one second.

lumbers right along with him, as long as he travels at a constant velocity (meaning in a straight line at a constant speed). But nearly any change in his velocity will, according to Einstein, produce a faint ripple in the field, called a gravitational wave. And if he undergoes a continuous sequence of velocity changes, such as by running up and down the stairs or going into orbit (not a wholly unpleasant thought), the result should be not a single gravitational wave but a chain of them.

Since gravity is a warping of space-time (see "'Round and 'Round"), and since any fluctuation in gravity is a distortion of that warping, a gravitational wave would produce some eerie effects as it rippled through town. Objects would be seen to move back and forth rhythmically, even though no force had been applied to them. This effect would go unnoticed by almost everyone, but all the same it would be exciting, humbling, and a fundamental validation of relativity, and Sheldon only pretends not to care with his airy "How does this gravity wave breakthrough help the man on the street?"[2] (When it comes to gravitational attraction, Sheldon's a regular Romantic poet. Does his mother not call him "Shelley"? Does he not sigh, "Ah, gravity, thou art a heartless Bysshe"?[3])

In many ways, gravitational attraction is like other kinds of attraction: it grows stronger with decreasing distance. For instance, no matter how close Penny already is, Kurt can increase his pull on her (gravitational or otherwise) just by bringing her a little closer. But when does that end? Won't she get to a point where his pull is so strong that nothing, not even light, can break free? And at $r = 0$, when she's actually *touching* him, why isn't his attraction infinite? Once two things are in physical contact, how can they ever break free of each other?

We may be able to sidestep this question, because it's not clear that it can ever happen. As r gets very tiny, reality gets very grainy, and it may be that it's physically impossible for two distinct objects to be separated by a distance of zero. String theory supposes that if two points are closer together than a certain minimum distance, there's no physical process that could ever distinguish between them, meaning that anything occupying either point would be seen as occupying the other as well. (It's like the game of tic-tac-toe, in which every **X** or **O** is understood to occupy its entire square despite physically covering only a portion of it.) If that's true, then as long as two objects remain distinct, they can never get closer together than some minimum distance.

Busy Prepositions

We think of gravity as pointing very definitely in one direction — up is up and down is down — but that's just our parochial, Earth's-surface–centric view. A less prejudicial terminology like *toward* and *away from*, or *in* and *out* might help emphasize gravity's omnidirectional nature.
The very use of the letter r (for radius), rather than the d (for distance) you might have been expecting, emphasizes the spherical symmetry of the $1/r^2$ laws.

But in the case of two people, the answer is much simpler: string theory or no string theory, they can never actually touch.

It's not unusual to have a law that works over enormous distances yet doesn't apply at short range. The $1/r$ model of ant spreading doesn't apply close to the mouth of the anthill, where the ants are climbing all over one another and haven't started spreading out uniformly. And for atoms, which is what Penny and Kurt are made of, at short distances the $1/r^2$ model of gravitational attraction loses out to the electromagnetic force. As she closes the distance to him, the electromagnetic repulsion between their outermost atoms becomes far stronger than the gravitational attraction.

If she continues trying to bring their atoms closer together, his atoms will either retreat or force hers back, maintaining the gap. It's similar to the feeling she would get if she tried to push the north poles of two magnets together: the sensation that there was a squishy invisible ball resisting her.

So she's never truly *touching* Kurt. What she's feeling is the force his outermost atoms exert on hers — the force of electromagnetic repulsion.

Not the most repulsive thing about him.

arms more than two miles long. Any passing gravitational wave should cause a momentary submicroscopic change in the relative lengths of the arms.

There are actually two LIGO installations: one in Louisiana and another in Washington state. The wave's precise time of arrival at each installation would reveal its direction of travel and thus what part of the heavens to search for its source.

So far, LIGO hasn't detected anything that can definitely be pointed to as a gravitational wave, but that's to be expected. All the objects we know of that are big and violent enough to produce detectable waves are pretty far away, and that $1/_r$ law is a killer. But who knows: the next time Kurt whips out his massive ego, be prepared for alarms to go off in Louisiana and Washington.

[SCIENCE TO COME]

$\mu G \neq 0G$

We hear from time to time about Howard's "zero gravity" experiences, first while training in an airplane in free fall (known casually as the "vomit comet") and later aboard the International Space Station. But the truth is that "zero gravity" is one of those dated phrases (like "outer space" and "rocket ship") that live on in science fiction but not in science.

Outer space is "space." A rocket ship is a "spacecraft." And Howard can never experience zero gravity. No human being can.

For one thing, although the Earth's gravitational pull diminishes rapidly with distance, at the altitude of the International Space Station it will only have fallen off by about 15%. At the altitude of the Moon it's almost 4,000 times weaker than at ground level, but that's still enough to keep the Moon falling around the Earth in its orbit, as Newton realized (see "KISS and Tell"). If the ISS, or the orbiting Moon, were to grind suddenly to a halt, it wouldn't continue to float through a tranquil sky; it would plummet at once toward Earth, bringing the astronauts and their flags and candy wrappers and everything else with it.

Howard can swim through the ISS with ease only because both he and it are in orbit around the Earth, and therefore his surroundings are falling around the planet exactly as fast as he himself is. The same thing happens to you in a roller coaster careening downhill. You're falling, and all your internal organs are falling just as fast. They're still right there inside you, but you can no longer feel them hanging down from your skeleton, so you feel "weightless."

Every piece of matter you can see (as well as many that you can't) exerts a gravitational pull on you, no matter how tiny. In order to experience no net gravitational pull whatsoever, you'd have to take yourself infinitely far away from everything, or else you'd have to surround yourself with an exactly balanced distribution of matter, so that a pull in any direction would be exactly balanced by a pull in the opposite direction.

That never happens. Even in orbit, Howard is constantly being subjected to itty-bitty tugs in various directions, and the more precise term for his almost-weightless condition is "microgravity." It's simply impossible for him, or for any astronaut, to escape gravity's influence entirely.

It would be easier to escape his mother.

1. "The Cooper-Hofstadter Polarization" (Season 1, Episode 9)
2. "The Relationship Diremption" (Season 7, Episode 20)
3. "The Big Bran Hypothesis" (Season 1, Episode 2)

SEVEN
FIZZ-ICS

Halo can take ten hours or more to play, *Battlestar Galactica* can take ten *days* or more to watch, but dropping a candy into a soda bottle yields ten tons of fun in mere seconds. And unlike *Halo* (a science-fiction video game franchise) and *Battlestar Galactica* (a science-fiction television franchise), Diet Coke and Mentos is a do-it-yourself experiment firmly rooted in science-*non*fiction:

✱ Step one: Drop a few Mentos candies into a freshly opened bottle of Diet Coke or other soda.

✳ Step two: Clean up.

The moment the Mentos hit the soda, a huge jet of foam surges out of the bottle. If conditions are just right, the eruption can reach a height of fifteen feet or more.

Mentos A brand name of mint candies distributed by Perfetti Van Melle. As with sheep or GloFish or faux pas, Mentos is both the singular and the plural form.

This spectacular effect is not the result of a chemical reaction between the candy and the soda. Chemically, there's nothing new being created; everything that fuels the geyser was there to begin with. It's simply being released all at once — an example of a physical change.

physical change A change to the physical form of a substance but not to its chemical composition, as when ice melts, coffee is ground, water and dirt make mud, or carbon crystallizes into diamonds.
chemical change A change to the chemical composition of a substance, as when vinegar and baking soda combine to produce foam, crude oil is refined into gasoline, or fenders rust.

Carbon dioxide (CO_2) is the gas that gives soda its fizz. Inside a sealed bottle, the CO_2 stays dissolved, floating around in bubbles too small to see. The built-up pressure keeps them from merging into larger bubbles. That's why an unopened bottle of soda looks flat.

Opening the bottle releases the pressure, allowing those tiny bubbles to begin clumping together into larger bubbles, which float upward. Dropping in a Mentos makes this happen much faster and very messily. The more Mentos, the faster the fizzing and the bigger the mess. But how?

If you peer into a glass of soda, you can see steady streams of tiny bubbles appearing seemingly from nowhere and rising toward the surface. They're actually forming at the edges of virtually invisible scratches and imperfections in the glass. In fact, manufacturers will often etch a pattern of deliberate scratches into the bottom of a beer glass or mold a raspberry-shaped glass bead into the bowl of a champagne flute to create multiple nucleation sites and encourage the formation of visually attractive bubbles.

nucleation The coalescing of small bubbles (or droplets) into larger ones, usually around an imperfection or a foreign object.

A soap bubble floating through the air has an approximately spherical shape. But the moment it lands on a wet bar of soap and its underside is no longer exposed to the air, surface tension flattens it out into a dome. For the same reason, when a microscopic bubble of dissolved carbon dioxide in a soda happens to brush against a microcrater or microcanyon in the glass surface, it jams itself into the imperfection and flattens out.

Additional passing microbubbles will crowd themselves into the same space, coalescing into a single large bubble that eventually becomes big and buoyant enough to detach from its nucleation site. The bubble begins rising toward the surface of the liquid. Along the way, it takes on more CO_2 and becomes more buoyant still, and by the time it reaches the surface it's large and moving swiftly. Meanwhile, down below, another bubble has begun forming, and the process repeats.

To the naked eye, Mentos candies appear to have a smooth, almost shiny coating, rather like oversized M&Ms. Under the microscope, however, that coating shows itself to be pockmarked with minuscule nooks and crannies, just the right size for a microscopic bubble of dissolved CO_2 to get itself lodged into. In fact, the rough surface of the Mentos is so good at pulling carbon dioxide out of solution that the bubbles it forms are big and traveling fast from the beginning. They don't just leisurely wander up and out of the bottle; they rush out in a torrent. Any liquid that's in the way gets carried out with them — sometimes as much as half the contents of the bottle.

You can demonstrate that this is a physical reaction and not a chemical one by first pouring the soda into an empty bottle. When you then add the Mentos, the effect is greatly diminished. Same ingredients, but a much less impressive eruption. That's because pouring the soda disturbs it enough to allow much of the carbon dioxide to float to the surface and fizz harmlessly into the atmosphere. By the time the first Mentos appears on the scene, the remaining dissolved CO_2 isn't generally enough to make much of a spectacle. (This also explains why gorging on Mentos and chugging soda — although we certainly don't recommend it — rarely leads to projectile vomiting, much less a ruptured stomach. By the time the soda meets the Mentos somewhere inside you, most of that gas has escaped.)

A New Kind of Gas-Powered Car

The entertainment company EepyBird is widely known for their eye-catching demonstrations of the Diet-Coke-and-Mentos effect. They've staged effervescent ballets involving hundreds, even thousands, of bottles of soda.

In 2010 they created a viral video featuring a "rocket car": essentially the front half of a mountain bike welded to a small utility trailer. Neatly arrayed on the trailer were 108 two-liter soda bottles containing not Diet Coke but Coke Zero (a nearly identical product in a more manly wrapper). Each bottle was preloaded with six Mentos and connected to a rear-facing nozzle fashioned out of sprinkler pipe. The "engine" was started via an intricate string-and-pin arrangement that released all 648 Mentos simultaneously.

Though the initial impulse lasted only a few seconds, the car ultimately rolled more than 220 feet ($1/24$ of a mile, or $1/3$ of a furlong), dribbling a distinctive Coke Zero creek in its wake. A year later, using a lighter-weight car fueled by only half as many bottles, EepyBird broke their own distance record.

Rumors that North Korea is working on a cruise missile powered by 7UP and Pop Rocks have not been confirmed.

---

* The filmmakers made a deliberate decision to substitute Coke Zero for Diet Coke. Market research suggests that men find the word "zero" more masculine than the word "diet." Market research also suggests that women find the concept of a soda-powered rocket car pretty stupid.

It's not known who first observed the Mentos-and-soda phenomenon, nor who first booby-trapped a roommate by hiding a Mentos under the cap of a full soda bottle. But once you've experienced it, it's natural to wonder about things like:

* Which sodas foam up most dramatically?
* Do some Mentos flavors work better than others?
* What's the most effective delivery system for getting the Mentos pellet into the soda?
* What's better: a quick, explosive eruption that leaves the bottle half full or a more gradual reaction that nearly empties it?

Questions like these are the reason there will always be a need for experimental scientists.[*]

Dr. Tonya Shea Coffey of Appalachian State University in North Carolina organized her sophomore physics students to investigate the underlying mechanics of this reaction. Their results were published in the *American Journal of Physics* — complete with data tables, photos from the scanning electron microscope, and detailed analysis. The young researchers took the problem quite seriously, controlling for an impressive number of factors, including speed of pellet delivery, spray angle, soda manufacturer, and expiration date. They also came up with a highly creative array of parameters to tweak:

* Liquid: soda, seltzer, or tonic water; diet or sugar-based; with or without caffeine
* Pellet: Mint Mentos, Fruit Mentos, Wrigley's Life Savers candies, and non-candies including sand, silicate beads, rock salt, and baking soda
* Pellet preprocessing: intact or crushed
* Temperature: room, prewarmed, or precooled

As expected, their results confirmed the importance

---

[*] During a private moment near the end of an emotionally draining day, Raj and Stuart share a nightcap in the comic book store. Unnoted but plainly visible by the cash register is a tube of Mentos. It's obvious that the two of them are ready for a little experimentation.[1]

of pellet configuration and delivery method. The more rapidly the pellet sinks, the more dissolved CO_2 it contacts per unit time, and the lower the chance of its premature ejection from the bottle. Video analysis of intact Mentos showed them plummeting through the liquid. Crushing them beforehand yielded a powder that sifted down far more slowly, giving a much weaker effect despite the far greater surface area.

The students also found that diet sodas gush farther than do sodas containing sugar. The most likely explanation is that sugar water, having a higher surface tension, is more resistant to bubble formation. (This is fortuitous, because the residue from sugar-based sodas is gummier and harder to clean up. Sticky sugar also attracts vermin, which is an important consideration if your cleaning-up skills resemble, say, those of a sophomore physics student.[2])

Looking for the Answers to Questions Nobody's Asked

Does it seem silly that someone would do all that Mentos research? Without a practical goal in mind, were the students just wasting snack foods?

That's the difference between pure research (the acquisition of knowledge for its own sake) and applied research (done with a further, generally more practical, goal in mind). What's important is the knowledge that's gained, not the motivation behind it. Someday in the future, someone might find a truly practical use for the Mentos rocket. When that day comes, the more already known about how and why it works, the better.

One unexpected result was that Mentos outperformed Life Savers, which appear far rougher under the microscope and thus should provide far more CO_2 bubble growth sites. The reason turned out to be the gum arabic present in the Mentos coating and not found in Life Savers. Gum arabic raises the bubble-forming rate by lowering the surface tension of the soda.

Another surprise was that even though Fruit Mentos have a coating that appears far shinier than that of Mint Mentos, they gave better results. The microscope revealed numerous gaps in the fruit coating, barely visible, through which the especially rough interior was exposed. Furthermore, the fruit coating dissolves more quickly than the mint coating.

Many objects can be dropped into a bottle of fizzy liquid to make it gush out, but — whether by design or by accident — Mentos combine an impressive number of advantages in one inexpensive package. There's a Nobel waiting for the person who develops a more effective soda displacement pellet.

But in the meantime, there's evidently more behind a simple "drop some Mentos in Diet Coke" than Penny realizes. Just working out the materials list could take the guys all evening.

Snap, Crackle, Pow!

Caltech's ebullient Chris Brennan positively effervesces when he talks about his research into cavitation. That's the formation of cavities (bubbles of gas or vapor) in liquids. Though desirable in sodas and bath soaps, it's a hazard to propellers and pumps.

A bubble being squeezed by the fluid medium that surrounds it stores up the energy of compression and heats up. As the pressure increases, the bubble may suddenly rupture or collapse, causing a jet of fluid to rocket into its interior with great force, releasing shock waves and heat. This process can be used for stripping off dead skin, shattering kidney stones, and mixing paints. But if not controlled, the cumulative effects of a series of violently imploding bubbles can be powerful enough to erode metal.

Just one more way soda's bad for your teeth.

IN WHAT UNIVERSE?

Mapmaker, Mapmaker, Make Me a Map

Maybe the reason we're having difficulty finding Leonard and Sheldon's apartment is that we're using the wrong system of street addresses. In North America, buildings along a street are commonly numbered sequentially, with odd and even numbers taking opposite sides of the road and the hundreds digit going up every block or every twentieth of a mile. But in parts of the UK, as well as other countries that were settled many centuries before anybody ever really needed to send junk mail to anybody else, houses are numbered chronologically, more or less according to the order in which they were built. Letter carriers and knowledgeable locals carry these non-intuitive mappings in their heads; everyone else is left to wander the streets, peering at every doorpost.

Is there a similar numbering system at work on *The Big Bang Theory*? (There's circumstantial evidence to support this hypothesis: the street number in the return address on Bernadette's wedding invitations is 720, yet the building she lives in has a large **3627** by the front door.[3]) Could it be that "2311 North Los Robles Avenue" refers to the 2,311th building constructed on North Los Robles Avenue?

Regrettably, this turns out not to be the case, as a quick trip to the Pasadena Department of Planning & Community Development confirms. (Or it presumably would, if one could get past the clerk's frosty "You expect me to spend my lunch hour helping you look up *what*?") In the 140-year history of the city, considerably fewer than 2,000 buildings have as yet been built along that fine thoroughfare. Anyway, the number **2311** is clearly visible on the outside of the building.[4]

Perhaps we're simply putting too much faith in the word "North." On the odd-numbered side of any "North" street in Pasadena, higher-numbered buildings are found to the right of lower-numbered buildings. Yet on *The Big Bang Theory*, the building to the right of number 2311 is number 2309 (the Urban Lights store).[5] That's the ordering that would be found on the odd-numbered side of a "South" street. And as it happens, South Los Robles Avenue — unlike North

Los Robles Avenue — does have a 2300 block; in fact, it's the last block on that street. But it's in San Marino, not Pasadena: one city too far to the south.

Then again, what the characters refer to as "North Los Robles Avenue" might not be the street the rest of the world refers to as "North Los Robles Avenue." Maybe they're using a secret street-renaming code to mislead eavesdropping invaders. After all, at one point Howard tells Leonard, "Turn left on Lake Street," and much later Sheldon refers to "the Coffee Bean over on Lake Street," although there is no Lake Street in Pasadena. Presumably they mean Lake Avenue, an even busier street than Los Robles and only a quarter of a mile to the east. Or maybe "Lake Street" is some other street's Klingon name.[6]

Is there *no one* we can trust to give us something as simple as a street address?

---

1. "The Date Night Variable" (Season 6, Episode 1)
2. Tonya Shea Coffey, "Diet Coke and Mentos: What is really behind this physical reaction?," *American Journal of Physics* 76, no. 6 (2008): 551–557.
3. "The Vacation Solution" (Season 5, Episode 16) and "The Vengeance Formulation" (Season 3, Episode 9), respectively.
4. "The Mommy Observation" (Season 7, Episode 18)
5. "The Spaghetti Catalyst" (Season 3, Episode 20)
6. "Pilot" (Season 1, Episode 1) and "The Decoupling Fluctuation" (Season 6, Episode 2), respectively.

EIGHT

NUH UH!

> Sheldon: So, if there's an earthquake and the three of us are trapped here, we could be out of food by tomorrow afternoon.
>
> . . .
>
> Leonard: Penny, if you promise not to chew the flesh off our bones while we sleep, you can stay.
> Penny: What?
> Sheldon: He's engaging in reductio ad absurdum. It's the logical fallacy of extending someone's argument to ridiculous proportions and then criticizing the result, and I do not appreciate it.
> — "The Dumpling Paradox" (Season I, Episode 7)

Trust Sheldon to say in thirty words what most people would say in one or two (unprintable) ones. And reductio ad absurdum isn't even the right term for what Leonard's doing. For one thing, he's exaggerating what Sheldon has said, not reducing it. For another, reductio ad absurdum isn't a logical fallacy at all; it's a logical analysis technique. But as long as Sheldon has introduced the subject of logic and logical fallacies, let's go there. To some it'll sound as though we've abandoned science for philosophy, but actually we're just shifting our focus from *what* we know to *how* we know.

Every one of us is the caretaker of a highly personalized mental representation of reality. No two representations agree precisely; each is an approximation based on an individual's unique set of experiences. Our models are constantly evolving, with our thought processes providing a framework for incorporating new perceptions and beliefs, and while the framework may be rigorous, the model itself rests on a vaporous foundation.

logical fallacy An attempt to prove something by using faulty logic, intentionally or not. If the resulting conclusion turns out to be true (or at least reasonable), a logical fallacy will often go undetected or unchallenged. An example is post hoc ergo propter hoc, the notion that because B happens after A happens, A must have caused B (see "Coming to Think of It"). Even if A doesn't cause B, it's often hard to prove that fact, especially if A is a rare event. Faulty logic doesn't necessarily give wrong answers, just unsupported ones. But since the right answers it sometimes gives can lead to a false sense of security, in many ways it's worse than no logic at all.

Many people are taught at a young age that facts are provable beliefs and opinions are unprovable ones. Yet in some sense both are subjective. What's factual to one individual might strike another as only opinion, if not dead wrong, and an opinion doesn't become a fact just by virtue of its being widely held. The philosopher John Stuart Mill noted, "If all mankind minus one, were of one opinion, and only one person were of the contrary opinion, mankind would be no more justified in silencing that one person, than he . . . would be justified in silencing mankind."[1] Even when all mankind is in agreement, it may be that a fact is nothing more than an opinion that has not yet been refuted (see "Past Performance Is No Guarantee"). New information, new thinking, or a change in the workings of the Universe might one day disprove any currently held "fact." At heart, everything is faith.

Sheldon grasps at this straw during the Physics Bowl competition.[2] Informed by Dr. Gablehauser that "the answer your teammate gave was correct," he pouts, "That's your opinion." (The tables are turned when Sheldon informs his mother, "Evolution isn't an opinion; it's fact," and she retorts, "And that is your opinion."[3] A maddening point from a scientific perspective, but irrefutable from a philosophical one.)

Despite the ultimate unprovability of whatever "facts" underpin any particular mental model, the model itself can still have an internal consistency. Having chosen (however arbitrarily) our beliefs, we can build up additional knowledge just by manipulating them according to the rules of logic. But those rules are strict, and manipulations that violate them (called logical fallacies) must be avoided. Misperceptions and mistaken assumptions may cause the model to part ways with reality, but logical fallacies invite the model to part ways with itself.

Reductio ad absurdum (Latin for "a reduction to the absurd") is not a logical fallacy. It's a specific deductive reasoning technique for using logic to prove the truth (or falsity) of a statement. Generally, it involves temporarily assuming that some doubtful statement is true and then using logic to conclude that true equals false. Since that statement is absurd, and since we made only one assumption, we're forced to conclude that the assumption itself must have been false.

reductio ad absurdum

If someone draws a square and says, "Look, I drew a triangle," here's how you can use reductio ad absurdum (politely, of course) to prove that the statement is false.

You might say: "We both agree that all triangles have three sides. We also agree that you've drawn a shape with four sides.

"Let's temporarily assume that your statement is correct: that the shape you drew is a triangle. If so, then that's the same as saying: 'This four-sided shape, being a triangle, has three sides,' or, more simply, 'This shape has both four sides and three sides.' That would be true only if four equals three — which is false. (In fact, it's absurd.)

"Since our conclusion is false, we must have either relied on a logical fallacy or included at least one false statement somewhere along the way. We didn't rely on any logical fallacies, and we know that all the statements we used are true, with the exception of 'This shape is a triangle'; that's the one we only *assumed* to be true. So that must be the statement that's false. Hence your shape is *not* a triangle (... *idiot*)."

This method only works if we make exactly *one* assumption; if we make more than one, how will we know which of them is (or are) false? Also remember that the word "absurd" refers only to the conclusion — not to your friend's knowledge of geometry, his drawing ability, or your choice of friends.

Reductio ad absurdum surfaces in fields as distinct as logical discourse and mathematics and was a favorite of the ancient Greek philosophers. It sidesteps wishful thinking, because regardless of what your intuition tells you, if your conclusion is absurd, then you either used faulty logic or started from faulty premises. (See how to prove that Socrates was mortal in "Coming to Think of It.")

Sheldon would do better to accuse Leonard of intentionally constructing a straw man argument. That's a particular kind of logical fallacy in which a person tries to disprove an idea by attacking a related idea without acknowledging the important differences.

For instance, when Howard proposes to enroll in

Sheldon's class, Sheldon proceeds to grill him on a slew of esoteric arcana in a frantic attempt to prove his own intellectual superiority. In so doing, he's treating Howard's reasonable claim ("I'm more than smart enough to take your class") as though it were the significantly different "I'm more than smart enough to *teach* your class." Howard's disgusted comeback ("Asking me a bunch of questions about a topic I'm not familiar with doesn't prove anything") indicates that he's spotted the straw man.[4]

The origin of the term "straw man" is unclear, but it probably has nothing to do with grasping at straws. For many, it calls to mind an army of scarecrows: as imposing as real soldiers at first glance, but on closer inspection, insubstantial and powerless to defend themselves.

It's easy to get away with a straw man argument if it's superficially similar to a legitimate concern and raises strong emotions. An example is when Sheldon sarcastically praises his mother's religious group for having recently become "willing to sail out into the ocean without fear of falling off the edge."[5] That's a straw man fallacy not because the world isn't flat, nor because such a notion deserves Sheldon's derision, but simply because while purporting to challenge the beliefs of Mary's group, he's scorning one that they don't in fact hold. They've never said they think the world is flat, so it's unfair of him to attack them as though they have.

Græcum mihi

If the concept of reductio ad absurdum comes to us from classical Greek philosophy, why do we use a Latin name for it? After all, Greek was good enough for *philosophy*, and for *paradox* and *dichotomy*, too.

The answer may be that in the intervening two-plus millennia, after Rome had taken over being Top Nation, Latin became the universal language for spreading knowledge. Many individual Greek words persisted virtually unchanged — *metaphysics*, *dilemma*, dilithium (or no — wait — that last one's from *Star Trek*) — but multi-word phrases often became better known in translation.

Apart from *philosophy*, you can still hear the echoes of ancient Greece in science-related nouns like *thermometer*, *television*, and *sociopath*. Which is remarkable because those are all things the ancient Greeks didn't have. Well, except sociopaths, maybe.

According to Sheldon, allowing Penny to sleep over runs the risk of stretching the apartment's earthquake supplies too thin. Rather than disputing this plausible but hardly catastrophic scenario, Leonard exaggerates it by rephrasing it to say that Penny may become a ravenous cannibal during the night. He backs up his straw man by deliberately setting up a false dichotomy: unless Penny sleeps elsewhere, cannibalism is likely to result.

(c) None of the Above

A false dichotomy is a logical fallacy that incorrectly implies that there are only two possible answers to a question. Whether disingenuously or accidentally, any possibility of an additional alternative, or tertium quid, is ignored.

Penny, wallowing in self-loathing, sets up a false dichotomy when she calls herself either "just an idiot who picks giant losers" or else one of those who "pick good guys, but turn them into losers." When Leonard denies that either of these is true, she sniffs, "Well, it's got to be one or the other; which is it?"[6]

Consider these additional examples of potentially unfair "either/or" thinking:

- "Either you're with us or you're against us."
- "If you're not part of the solution, you're part of the problem."
- Sheldon: "The friend of my enemy's girlfriend is my enemy."[7]
- "Are you lying, or are you just stupid?"
- "Who had more Facebook friends: Adam or Eve?"

Oddly, Leonard's statement is unquestionably true. Penny *could* become a ravenous cannibal during the night, whether or not there's an earthquake and whether or not the supplies run out. In fact, theoretically she could become a cannibal at any moment (whether or not she sleeps over at all), just as Leonard could spontaneously combust or Sheldon could sprout a pair of antlers.

But it's a remote possibility — an absurdly remote one (which may account for Sheldon's confusion about the word *absurdum*) — and most of us choose to live our lives with virtually no thought to the possibility that our friends might suddenly take a walk on the anthropophagous side. Still, it's good to keep an eye on your friends. And if they're sleeping over, make sure they have easy access to the fridge.

400 pages to lay the groundwork for a proof that 1 + 1 = 2), it follows a structure that rigorously extends the existing foundation, brick upon brick. This is necessary because a foundation is only as strong as its weakest spot.

Apostol co-created *Project MATHEMATICS!*, a collection of video lessons covering the basics of geometry and trigonometry at a high-school level. One of the most popular segments in the series speculates on the geometrical methods behind an amazing engineering achievement of ancient Greece: the excavation of a tunnel two-thirds of a mile long by two teams of miners who started from opposite sides of a mountain and met deep in the middle.

He's long been a champion of the "visual calculus" developed by astrophysicist Mamikon A. Mnatsakanian, an extraordinarily intuitive way of solving complex mathematical problems simply by visualizing the ways in which line segments "sweep out" regions of space as they travel along curved paths.

OUT TO LANDS BEYOND

"[Writing poetry is] like doing a scientific experiment. It's mostly failure before you get anything."

Jessica Goodfellow (Caltech M.S. '89) is a writer of poetry and a four-time Pushcart Prize nominee. With a bachelor's in economics from Brigham Young University, she began working toward a Ph.D. from Caltech. She got as far as completing her master's degree in social science and publishing a working paper entitled "An Experimental Examination of the Simultaneous Determination of Input Prices and Output Prices" before deciding to pursue her first love: creative writing.

Goodfellow's writing probes the ageless struggle to reconcile faith and science. It's easy to see the influence of both in her hauntingly introspective poems. "A Pilgrim's Guide to Chaos in the Heartland," about humans' longing to find patterns in randomness, is peppered with random digits that crop up between the syllables like poppy seeds in a muffin. "Navigating by the Light of a Minor Planet" opens with the observation: "The trouble with belief in endlessness is / it requires a belief in beginninglessness."

Garrison Keillor featured her poems "The Invention of Fractions" and "In Praise of Imperfect Love" on NPR's *The Writer's Almanac.*

1. John Stuart Mill, *On Liberty*, 1869.
2. "The Bat Jar Conjecture" (Season 1, Episode 13)
3. "The Electric Can Opener Fluctuation" (Season 3, Episode 1)
4. "The Junior Professor Solution" (Season 8, Episode 2)
5. "The Rhinitis Revelation" (Season 5, Episode 6)
6. "The Tangerine Factor" (Season 1, Episode 17)
7. "The Deception Verification" (Season 7, Episode 2)

NINE
BETTER LIGHTING THROUGH CHEMISTRY

> Leonard: Why would you want a glow-in-the-dark ant farm?
> Sheldon: They do some of their best work at night.
> — "The Hot Troll Deviation" (Season 4, Episode 4)

Sheldon's right: ant colonies don't follow the standard daytime/nighttime sleep schedule human colonies do. Though their activities may shift after sunset due to the drop in illumination and temperature, they're still awake and bustling long after most two- and four-legged creatures have gone to bed. He's also not making up the notion of an ant farm with glow-in-the-dark sand. That's a reality, although the phosphorescence that produces the glow fades after about fifteen minutes.

Some ant farms avoid the fade-out problem by replacing the sand with an ant-friendly translucent gel and illuminating it from beneath. (Another solution, not yet available, would be for Amy Farrah Fowler to crossbreed an ant with a firefly. Picture a new species of insect miner with its own built-in headlamp — although most likely the lamp wouldn't be located on the *head* . . .)

> **phosphorescent** Able to "store up" light and then give it off for a time. Best viewed in the dark. Examples include plastic stars on bedroom ceilings, dots on the faces of some watches, and glow-in-the-dark toys.

Phosphorescence is sometimes confused with fluorescence. The two phenomena are distinct, though their underlying causes are related. Fluorescence typically looks like unrealistically bright coloration, and it requires illumination. Phosphorescence makes things glow in the dark after having been illuminated. Neither of these has to do with generating light, only with releasing previously acquired light. Between when the light is acquired and when it's released, it's stored in the electrons of the material's atoms.

> **fluorescent** Able to emit more of a given color than is present in the incoming illumination. Often quite striking under black-light illumination. Examples include highlighting pens, black-light posters, GloFish (see "Night Fishing"), and tonic water (but not club soda).

Every atom is surrounded by electrons, all constantly gaining and losing bits of energy. Whenever an electron absorbs or emits energy, its energy level changes. But the laws of the Universe restrict it to only certain specific energy levels, depending on the type of atom it's in. No electron is ever found at an energy level that its atom doesn't permit, not even for an instant. It's as if your car could go exactly fifty miles per hour, or forty or twenty, but never thirty or twenty-one or thirty-eight and a half. As you pressed down on the gas pedal at twenty miles per hour, for a while nothing would change — and then suddenly the car would be doing forty, without any kind of gradual transition. There wouldn't even be a moment where it was passing through twenty-five. It would be at twenty; then it would be at forty. Unpleasant for you, to be sure, and murder on your engine, but for electrons, instant switchovers like this are just the nature of things. There's no partway; no in-between.

This isn't as extraordinary as you might think. There are plenty of things in this world that you can't have an in-between amount of — holes, paper cuts, crumbs (as comedian George Carlin observed, "You break a crumb in half, you don't have two 'half-a-crumbs' — you got *two crumbs,* man!"[1]). And who ever heard of two *and a half* men? Impossible! (Or at least really, really gross.) As for why things are this way, it has to do with resonances and frequency restrictions. Resonances are the reason a bugle can produce only certain notes of the scale and none of the notes in between, the reason a washing machine shakes violently only at certain spin speeds and not others, the reason it's impossible to get a playground swing to go just slightly faster or slower.

An electron doesn't have a gas pedal. It gets boosted up to a higher energy level by absorbing a photon (a packet of light), and when it drops to a lower energy level, it emits a photon. The energy contained in either of those photons equals the energy difference between the electron's "before" and "after" levels. What we think of as the colors of the rainbow are produced by photons containing different amounts of energy, with red-light photons having less energy than violet-light photons.

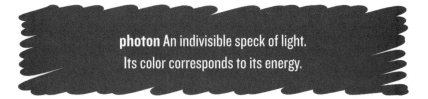

photon An indivisible speck of light. Its color corresponds to its energy.

Since an atom's electrons are restricted to only a certain set of energy levels, and since the color of a photon corresponds to its energy, the only light an atom can absorb is light whose color matches the difference between two of the energy levels that are permitted to that atom's electrons. A photon of incoming light of just the right color can kick an electron up into a higher level, while photons whose energy doesn't correspond to an allowed increase in energy level aren't absorbed and just continue on their merry way.

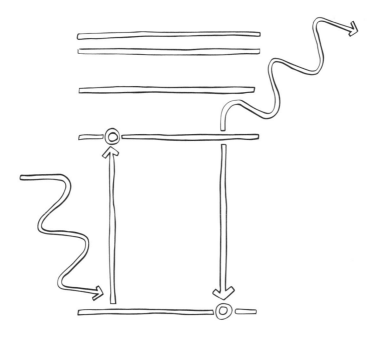

Blue light being absorbed and reemitted. The vertical dimension shows the electron's energy, not its position. Horizontal lines indicate a few of the energy levels permitted to it.

The colors an atom can emit are the same as those it can absorb. If an atom absorbs an incoming photon of blue light, kicking an electron up into the next higher energy state, then as the electron drops back down again, the atom will emit that same color of blue. Under a yellow light, such as a streetlight, such an atom may appear black; with no blue in the incoming light, there's nothing for the atom's electrons to absorb and then emit.

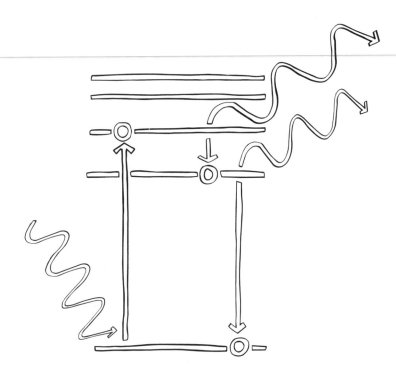

Violet light being absorbed; orange and blue light being emitted. The electron skips a level on the way up and comes back down level by level.

But suppose the light striking the atom happens to correspond to a boost of two or more energy levels instead of just one. An electron absorbing one of these extra-energetic photons will skip one or more levels all at once. It would be as if that car that was doing twenty suddenly jumped not to forty but to fifty. And then? Regardless of what happens on the way up, an electron returning to lower levels typically drops only one level at a time. With each step down, it emits a photon whose color corresponds to the energy difference for that step.

If the electron skipped one or more steps on the way up, then none of the colors it emits on the way down will match the one it absorbed on the way up. This is fluorescence, and it effectively turns one color into two or more others. Because the colors it yields may not even have been present in the incoming light, it can lend objects an oddly vivid appearance, which is why many tropical fish fluoresce dramatically under a bright blue aquarium bulb.

The incoming light doesn't have to be in the visible range. Those gaudy posters from the 1960s were designed to look their best when illuminated by a "black light" bulb: a lamp that emits a large amount of ultraviolet light, which is invisible to the human eye. (The purple color of such bulbs is only a side effect of their real purpose.) Crushing sugar crystals also creates ultraviolet flashes, under which wintergreen oil fluoresces blue. This accounts for the dramatic effect you see when you look in a mirror in a dark room while biting down on a Wint-O-Green Life Savers candy with your mouth open, you slob.

In fact, the incoming energy doesn't even have to be supplied by light. Fluorescent light fixtures work (at least, after a moment or two of thinking and blinking) by scattering an invisible electron beam through a glass tube with a chemical coating on the inside. Energy from the electron beam boosts electrons in the coating into a higher energy state. As they return to their original state, one step at a time, they fluoresce bluish and yellowish.

Like fluorescence, phosphorescence requires an external energy source and comes from electrons returning to lower energy states. The difference is that it proceeds more slowly. Electrons usually emit energy almost as soon as they absorb it. They can't hold onto it for more than a fleeting instant before radiating it back out, which is why the moment the light goes off and all that energy stops flowing, you're plunged into darkness.

With phosphorescent materials, the process is the same but the timing's different. Owing to particulars of their atomic structure, their electrons take an appreciably longer time to emit the energy that they absorbed while the light was on. What you're seeing after the light goes off is a form of time-delayed re-radiation.

Since it takes a fairly bright light to get a glow-in-the-dark toy to glow even dimly, it may seem as though energy is somehow becoming lost inside the material, but that's not so. The electrons eventually emit all the radiation they've absorbed; they just do it a little bit at a time, over a longer time period. If you were to tally up all the energy going in and coming back out, you'd see that nothing is lost. This is an example of the law of conservation of energy, and if you felt like stating it formally, you could say that at thermal equilibrium the emissivity is the same as the absorptivity.* (But why stand on formality?)

---

* The law of conservation of energy is assumed to be true, and is allowed to be called a law, only because no one has yet been able to show a case in which it's false (see "Past Performance Is No Guarantee").

Surprisingly, the glow of the material called white phosphorus is due not to phosphorescence at all, but to a chemical reaction with the oxygen in the air. As for the genetics lab's "glow-in-the-dark bunny" mentioned by Howard, that's a mystery.[2]

Phosphorescence occurs in the light as well as in the dark, but naturally its glow is easier to see in the dark. It's also possible for fluorescence and phosphorescence to occur together, with electrons jumping up two or more energy levels at once and then gradually making their way back down.

With no source of new energy to "recharge" its electrons, the glow of a phosphorescent material tapers off over time. Should Sheldon become so interested in the nocturnal labors of his ants that he ends up watching them all night long, he'll find their sand fading out long before the sun fades in. He might be better off with fluorescent sand, perhaps illuminated not with eye-poking white light but with ultraviolet light or an invisible electron beam. On the other hand, a radiation bath like that might cause some freakish mutations in the ants' DNA, and the next thing you know, they're growing large enough to break out of their enclosure and terrorize Pasadena. (But not for long, because as he points out elsewhere, a giant ant would be crushed under the weight of its own exoskeleton.[3] Ew.)

Fade to black.

How Green *Was* My Valley, Actually?

Just as the incoming light that produces fluorescence or phosphorescence doesn't always have to be visible, neither does the emitted light. The infrared (three syllables: rhymes with "in the head," not "impaired") is a band of colors that, like the ultraviolet, happens to be invisible to the human eye. As the names suggest (to Latin speakers, anyway), the ultraviolet is "beyond" violet and the infrared is "below" red. The red heat lamps found above buffet tables and in bathroom ceilings produce infrared radiation. As with those purple "black light" bulbs, the visible color is only a side effect.

Through a process called solar-induced chlorophyll fluorescence, plants exposed to ordinary daylight give off a healthy infrared glow. We can't see this fluorescence, but an infrared detector can, and false-color satellite images of the Earth typically show areas of heavy vegetation in bright red (because that's the visible color normally chosen to stand in for invisible infrared).

At Caltech, chemistry professor Paul Wennberg is using the infrared signal of plants to measure global photosynthesis levels from space. This gives a much more accurate and unambiguous picture of plant and algae activity than our familiar shades of leafy green do.

And it can't be fooled by fake trees and AstroTurf.

A Nasty Glow to the Head

In Sir Arthur Conan Doyle's *The Hound of the Baskervilles*, Sherlock Holmes declares a glow-in-the-dark paint to have been made from "a cunning preparation" of phosphorus. Sheldon makes a similar assessment of the paint used by Howard and Raj one Halloween: "That one was clever. Skeleton with phosphorus on a zip line."[4] Sherlock and Sheldon were both undoubtedly wrong.

White phosphorus (the form of phosphorus that glows in the dark) is nasty stuff. It can ignite at room temperature, emitting a toxic smoke powerful enough to corrode metal and eat flesh away. (Doesn't sound all that cunning, does it?)

A hypothetical paint containing white phosphorus, assuming it didn't burst into flame the moment it was exposed to the air, would present serious health risks. Until the 1906 Berne Convention outlawed its use in the manufacture of matches, white phosphorus was responsible for a widespread and truly dreadful condition known as "phossy jaw," the bane of workers in match factories and fireworks plants. Airborne phosphorus particles lodging in the mouth invaded the mucous membranes; over time, the jawbone would take on a lovely greenish glow in the dark. Then it would rot away.

Again . . . ew.

---

1. George Carlin, "Some Werds," *Toledo Window Box* LP, 1974, Little David/Warner Bros. Records, Burbank, California.
2. "The 43 Peculiarity" (Season 6, Episode 8)
3. "The Wheaton Recurrence" (Season 3, Episode 19)
4. "The Good Guy Fluctuation" (Season 5, Episode 7)

TEN
SLITS AND STONES

> Sheldon: So if a photon is directed through a plane with two slits in it, and either slit is observed, it will not go through both slits. If it's unobserved, it will. However, if it's observed after it's left the plane but before it hits its target, it will not have gone through both slits.
>
> Leonard: Agreed. What's your point?
>
> Sheldon: There's no point. I just think it's a good idea for a T-shirt.
>
> — "Pilot" (Season I, Episode I)

Significantly, the very first words spoken on the series concern the one phenomenon that, according to Caltech physics Nobelist Richard Feynman, "is impossible, *absolutely* impossible, to explain in any classical way, and which has in it the heart of quantum mechanics. In reality, it contains the *only* mystery."[1]

The mystery hinges on the wave/particle duality of Nature, and it starts with a centuries-old question about what light's made of. Is light a shower of tiny particles, as Isaac Newton suspected? Is it a wave, as Dutch physicist Christiaan Huygens (pronounced like the first and last syllables of "highway wagons") proposed at around the same time? Is it some tertium quid (see "Nuh *Uh!*")? It bounces off things the way sound waves bounce off buildings, but it can travel through a vacuum the way sound waves can't.

> **wave/particle duality** The tendency of elementary particles (particles with no discernible internal structure) to behave sometimes like waves and sometimes like particles.

It's an important question because light is energy, and people like putting energy to work, and we can make better use of things if we know how they work. And for a long time, no one knew how light worked.

So let's talk about bricks for a minute. We've always known how *they* work. Picture a lamp burning near a brick wall. It's no surprise that the lamp sheds more light on the bricks closest to it than on those farther away. But why?

Suppose you had access to some type of super-special tile that heats up briefly whenever it absorbs energy. If you were to hit such a tile with a rock, or squirt it with a hose, or shine a strong flashlight at it, or shout at it, or let the ocean waves pound against it, it would heat up. Exactly how hot it got would depend on how hard you threw the rock, or how strong the jet of water was, or how powerful the flashlight or the shout or the ocean was.

Let's cover our brick wall with thousands of these magic tiles. The lamplight falling on them causes them to heat up — but not all to the same temperature. The closer a tile is to the lamp, the more light falls on it and the hotter it gets. If the lamp has a dimmer switch, we can turn down the brightness, but we'll still see that the closest tiles grow hottest.

Now imagine we've brought in Newton and Huygens (they aren't very busy these days) and asked them to explain why this is.

"Easy," says Newton. He stands next to our lamp and starts hurling handfuls of pebbles in random directions. Some of the pebbles miss the wall tiles, but others hit them. Each time a pebble hits a tile, the tile absorbs energy from the pebble and heats up a little.

"These pebbles are like the particles that light is made of," he explains. "I'm throwing them like this because that's what light does: it sprays out little particles in all directions."

Thanks to the $1/r^2$ law (see "The Gravity Situation"), the pebbles traveling outward from Newton's hand fan out in space like a dandelion going to seed. The tiles closest to him encounter the densest barrage of rocks and absorb the most energy, so they become the hottest.

Huygens shakes his head. "No rocks," he mutters. He begins to hum, loudly. The sound waves ripple out from him, and some of them reach the wall tiles. As the tiles absorb the energy of the waves, they heat up.

"These ripples of sound are like the waves that light is made of," he explains. "I'm sending them out like this because that's what light does: it fans out in all directions like a wave."

The waves traveling outward from Huygens trace out ever-widening circles, like ripples on a pond (see "Can You Hear Me Now?"). As they widen, their energy is spread more thinly, their intensity diminishes, and they grow fainter. The tiles closest to him encounter the most intense waves and absorb the most energy, so they become the hottest.

Let's simplify our three-dimensional experiment by making it somewhat more two-dimensional. We'll give each of our experimenters his own above-ground swimming pool to play in. Running along the far wall of each pool, just at the waterline, is a single row of those magic self-heating tiles, with their tops sticking out into the air and their lower halves submerged. Newton's pool is frozen so that he can skitter rocks across it. Huygens's pool isn't, so he can drop rocks into it and watch the surface ripples travel outward. They try out their pools, and again they both find that the closest tiles absorb the most energy and get the hottest. Now what?

We notice that if we cut a rectangular hole in a board and place it between our lamp and the wall, the hole casts a sort of reverse shadow on the wall. Now only the tiles within that reverse shadow will heat up; the other tiles, not receiving any light from the lamp, remain cool.

"Well, obviously," say Newton and Huygens, and each of them fashions a similar board, stands it in his pool, and shows how his model explains this shadow. The boards block most of the rocks on Newton's pool and most of the ripples in Huygens's, and only those rocks and ripples that happen to be heading directly for the hole are able to slip through.

Now imagine that the hole in our board is adjustable. We narrow it first to a slot, then to a slit, then to a still narrower slit. As the slit gets narrower, the reverse shadow gets narrower too (no surprise there). But after a while, although we're still shrinking the slit, the shadow begins to grow *wider.* It gradually fans out across the wall, making a much fainter version of the wide pattern we had when there was no slit at all.

What's happening is that the slit is diffracting the light, or spreading it out. Tiles farther out on the edges of the wall now see light shining through the slit. From their point of view, the slit is acting as a light source.

"Right," says Newton, "because with such a narrow slit, most of the particles that manage to squeeze through will ricochet wildly off its edges." And he confirms this by sliding a handful of rocks through the slit in his board. They fan out.

"No," says Huygens, "it's because when a wave passes through a narrow slit, it spreads out. The edges of the slit act like breakwaters, blocking neighboring ripples that would otherwise prevent the wave from spreading left and right. That sometimes happens to noises coming into a building from outside; the window spreads the sound out, and you can easily hear it no matter where in the room you're standing." And he demonstrates this by sending a wave through the slit in his board. It fans out.

"Light's a rock, Chris," glares Newton.

"Light's a ripple, Zack," glowers Huygens.

Which one of them has the right explanation? Is

there some way to distinguish between rocks and ripples? We'd better figure it out soon — they're both holding rocks.

One thing rocks can't do, but ripples can, is cancel each other out. Perhaps if we can make two beams of light overlap and then see whether they cancel each other out (the way waves do) or simply pile onto each other (the way rocks do), that will tell us whether light is a particle or a wave.

So we cut a second slit in our board alongside the first and invite Newton and Huygens to do the same. At long last, we have something that produces different results in their two setups. When Newton slides rocks through the two-slit setup, he gets side-by-side diffraction patterns. When Huygens sends waves through, he gets a beautiful and exotic interference pattern.

interference pattern A pattern formed when other patterns overlap in such a way that they alternately cancel and reinforce one another.

An interference pattern typically looks nothing like the patterns that produced it: in this case, the diffraction patterns generated by the two slits. (In fact, it more or less resembles the Doppler effect costume Sheldon wears to Penny's Halloween party later in the season, though it has nothing to do with the Doppler effect. Or Halloween.[2]) How hot each of Huygens's tiles gets depends on how far it is from each of the two slits. Wherever the relative distances

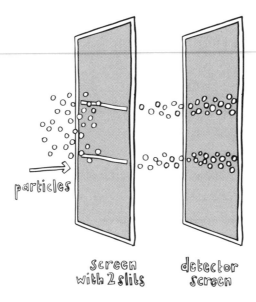

particles

screen
with 2 slits

detector
screen

The side-by-side diffraction patterns that result from putting
rocks through twin slits

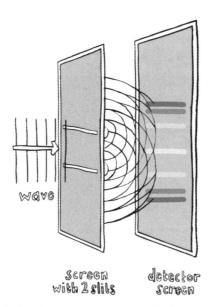

wave

screen
with 2 slits

detector
screen

The funky interference pattern that results from putting a wave
through twin slits

to the two slits happen to work out just right, a crest from one slit always arrives at the same time as a crest from the other slit, followed by a trough from one slit at the same time as a trough from the other slit. The resulting doubly high crests and doubly low troughs make for doubly hot tiles at these positions.

But wherever the relative distances to the slits are such that a crest from one slit always overlaps a trough from the other slit and vice versa, they'll cancel each other out, making no wave at all. Tiles at these positions remain cool.

The exact positions of the pattern's hot and cool spots depend on (among other things) the wavelength, which is the distance between successive wave crests. And if light is a wave, its wavelength should correspond to its color.

So let's shine a wide beam of monochromatic light on our side-by-side slits and see what happens. If light is a stream of particles, then we'd expect each particle to pass through one slit or the other (or else neither), just like a shower of itty-bitty rocks, and we'd see a pair of diffraction patterns on the wall, side by side. But if light is a wave, then we'd expect it to pass through both slits at once and spread out beyond them in two sets of widening ripples, producing a multi-banded interference pattern on the wall.

monochromatic Consisting of only one color of light.

In 1803, British scientist Thomas Young tried this experiment. (Actually, he used pinholes and sunlight instead of slits and monochromatic light, so the effect he got was smaller and fainter but more colorful and intricate.) And what did he see? A lovely interference pattern, suggesting very strongly that light is a wave. ("Told ya it's not rocky, fella," smirks Huygens to Newton. "It's ripply — believe it or not.")

Is he right? Let's go back to our original brick wall, only this time, instead of magic self-heating tiles, we'll pave it with an array of sensitive photodetectors — something Newton, Huygens, and Young were never able to do (perhaps owing to university budget restrictions). When we turn on our lamp, it illuminates the entire wall and all the photodetectors start triggering like mad. But if we grab the dimmer switch and start turning down the lamp, we eventually reach a point where the detectors appear to be triggering only at odd intervals, with those closest to the lamp triggering the most often. At this level of illumination our eyes can't make out any light at all, but the sensitive detectors still can, and they're acting as though they're detecting individual "specks" of light spattering the wall like raindrops. For our purposes these tiny specks of light (photons) are infinitesimal. No matter how compact our detectors are, nor how closely they're crammed together, no photon will ever overlap two detectors.

The energy of the photons corresponds to the color of the lamplight, but if we can continue to make the lamp fainter without changing its color, the photons won't become weaker or less energetic; they'll just arrive less often. (That's not so unusual: the raindrops don't get smaller just because the rainstorm is tapering off.) So it looks as though light's a stream of particles. But how does that account for what Young saw? (Now both Newton and Huygens are scratching their heads, and it'll take an Einstein — literally — to sort it all out.)

Let's insert the two-slit board into our apparatus and turn the lamp so low that it's sending out only one photon at a time. The light beam is wide enough to span both slits, and if photons really are like rocks, we can expect to see at least some of them coming through one slit or the other (more or less randomly) and gradually building up into two side-by-side diffraction patterns. We do indeed see the photons striking the wall one at a time, at unpredictable positions. But when we track those positions over time, they don't produce side-by-side diffraction patterns. Instead, they accumulate into an interference pattern: bands of many photons alternating with bands of almost no photons.

How can this be? Presumably, the interference pattern was produced by the overlapping of the two diffraction patterns. But that implies that each photon must somehow have gone through *both slits at once.* It's as though something that started out as a particle at the lamp, and ended as a particle at the wall, turned into a wave in between.[*]

This gets even stranger when the experiment is performed not with light but with a beam of things

---

[*] In a 2002 poll, the readers of *Physics World* magazine voted the double-slit experiment the most beautiful experiment of all time. Who says nerds aren't aesthetes?

that we "know" are particles: electrons, atoms, even large molecules. We can crank the intensity of the beam down so that only one of our tiny bullets hits the wall at a time, and we'll see each one fetching up at a random spot, as expected. But as long as the beam is broad enough to span both slits, the pattern of discrete hits on the wall will build up not into two side-by-side diffraction patterns but into one multi-banded interference pattern. It's as if each piece of matter becomes a wave, goes through both slits at once, and turns back into matter as it comes out the other side. (This realization is what fuels Sheldon's Cheesecake Factory epiphany: "I can't consider the electrons as particles. They move through the graphene as a wave. It's a wave!"[3])

Different Flavors of Light

In the particle model of light, the color is a manifestation of the energy carried by each particle, while the intensity is given by the number of particles arriving at a given area in a given amount of time.

In the wave model, the color depends on the light's wavelength (the distance from one crest to the next), while its intensity depends on its amplitude: higher crests and deeper troughs give brighter light.

The advantage of monochromatic light is that when the light is behaving like a stream of particles, all its photons will be at the same energy level, and when it's in a wave mood, all its waves will have the same wavelength.

Let's try to catch that intermediate wave sneaking through the two slits simultaneously. With light, this is easy to do: we just set up a pair of photon detectors right before the slits. When we turn them on, we're surprised (and perhaps a little annoyed) to find that the two detectors never trigger simultaneously. Each photon triggers only one or the other of them. And the pattern on the far wall has changed. All we see now are side-by-side diffraction patterns, just as we'd expect if light were made up of particles that could only pass through one slit or the other. It's as though the light "knows" that we're trying to catch it acting like a wave and insists on acting like a particle instead. Of course, it doesn't know anything of the kind, but we can't detect it without interacting with it, and we can't interact with it without affecting it.

This is the case even if we only use one detector, position it close to the wall, and don't even turn it on until after the photon has passed through the slits and it's too late for it to "change its mind." And "detector" doesn't mean "expensive machine we watch carefully"; it means *anything anywhere* that could *ever* be affected by which path was taken. No matter how we arrange the experiment, once we've added anything that could ever indicate which slit(s) an individual particle or wave has "chosen" to go through, the multi-banded interference pattern disappears.

Which brings us back to Sheldon's T-shirt idea. His description is a little careless, but what he's talking about is that you can't dissect a wave of light. The moment you try, it starts acting like a beam of particles instead.

In the words of Leonard, later in the pilot episode: "It's a paradox. And paradoxes are part of Nature."[4] But it's only a paradox because we picture particles and waves as different things. Stop thinking of particles as "little hard balls of stuff." Think of them as regions where multiple waves of particle-potential have combined to make a single wave of particle-ness. (See? Much more intuitive. It's *all* waves!)

Danish physicist Niels Bohr summed up this peculiarity of Nature in his complementarity principle of quantum mechanics: everything has both a particle nature and a wavelike nature, and which of the two is observed depends on the observer.

Then again, according to quantum mechanics pioneer John Wheeler, "Bohr used to say that if you aren't confused by quantum physics, then you haven't really understood it."[5]

Now *there's* a T-shirt!

EUREKA! @ CALTECH.EDU
The Zero-Slit Experiment

As demonstrated by a team that included Caltech theoretical chemist Vincent McKoy, it's possible to perform the electron version of the double-slit experiment without an electron beam and without any slits at all.

The electrons around an atom aren't all at the same distance from the nucleus. When two atoms pair up to form a molecule, they share some of their outermost electrons, but for the most part each atom keeps its innermost (core) electrons to itself.

By bouncing a photon of light off the molecule, however, it's possible to knock out one of these inner (core) electrons and send it sailing off toward a detector screen. If the two nuclei are different, or if they aren't the same distance from the source of light, then the electron is more likely to be ejected from one atom than from the other, and the detector pattern reflects this asymmetry. But if the atoms are identical, and if the molecule is held exactly perpendicular to the beam of light, then the ejected electrons build up into an interference pattern. The only conclusion we can draw is that the electron has come not from one atom or the other but from both at once.

It may help to think of atoms as being surrounded not by orbiting electrons but by clouds of probability that, when probed, appear as electrons (see "Says You!"). If two separate atoms' clouds were probed simultaneously, could they team up to coproduce one electron?

Evidently, that's exactly what they *are* doing. Like the double slits, they somehow act as partners in the role of creation. The photon strikes both atoms simultaneously, each of them emits *something*, and the two somethings somehow come together to form an electron.

Replace the word "atoms" in that sentence with the word "slits," change "photon" to "electron" or "electron" to "photon," and you've got the double-slit experiment.

In What Universe?

We Have a Visual

Whether or not the apartment building appears on any map, it ought to be visible from the sky. At the end of the season 5 closer, the overhead camera pulls back dramatically, taking us from the rooftop of the apartment building straight up to the altitude of "the Google satellite."* A quick finger on the pause button reveals that the building is located on the west side of a north-south street. In Pasadena, the west side is the odd-numbered side. Could we be looking down on 2311 North Los Robles Avenue?[6]

No. The building we see is in Pasadena, all right, but it appears to be located at approximately 215 South Madison Avenue, two blocks east of Los Robles. And the roof we see looks substantially different from the roof of the building that actually stands at that address.

Another dramatic pullback in an earlier episode pays homage to the *Star Trek* franchise.[7] It begins with a furious Sheldon bellowing Wil Wheaton's name into the cosmos and ends high above the planet. But since he's at Stuart's comic book store at the time, not at home, that's not much help. (As a matter of fact, freeze-framing shows that he's located in a building that has apparently been grafted onto the Hollywood & Highland Center, a tony Hollywood mall ten miles outside Pasadena.)

We could be forgiven for starting to suspect that this is some other Pasadena, California, on some other Earth in some other universe. Maybe the street names can give us a clue.

"Los Robles" means "The Oaks." It's one of a handful of oak-related street names in the Pasadena area, along with Oak Knoll, Oak Grove, Oakdale, Oakwood, Fair Oaks, Live Oaks, Pine Oak, and just plain Oak. Those Pasadenans do love their oaks; there's even a species native to the region called . . . the Pasadena oak.

"Madison" means "son of Matthew," and it'd be sweet if "Matthew" meant something like "we love oaks, too!" — but it doesn't.

Feels like we're starting to reach a little bit, doesn't it?

---

* Actually, Google doesn't own any satellites. Much of the imagery seen on Google Maps and Google Earth comes from standard aerial photography, where clouds are less of a nuisance.

1. Richard P. Feynman, *The Feynman Lectures on Physics*, vol. I (Reading, MA: Addison–Wesley, 1970), 37-1. http://feynmanlectures.caltech.edu/I_37.html#Ch37-S1 *shortcut:* http://DaveZobel.com/-bbfd
2. "The Middle Earth Paradigm" (Season 1, Episode 6)
3. "The Einstein Approximation" (Season 3, Episode 14)
4. "Pilot" (Season 1, Episode 1)
5. John Horgan, "Quantum philosophy," *Scientific American* 267, no. 1 (1992): 94–101.
6. "The Countdown Reflection" (Season 5, Episode 24)
7. "The Creepy Candy Coating Corollary" (Season 3, Episode 5)

ELEVEN
MAKING MEMORIES

Although Sheldon would more accurately call it a "game cartridge" than a "memory card," the object his mom included was indeed a core component of early gaming consoles such as the Nintendo 64. Everything a specific game needs — graphics, sound effects, logic — is built into the cartridge. To play a different game, the cartridge is simply popped out and a new one is popped in. Often, the game cartridge could also store the current state of the player's game; it's this feature that Sheldon intends to use so that he can pick up right where he left off nine years earlier.

Cartridges for the Nintendo 64 had a maximum capacity of sixty-four megabytes of storage. But the

"64" in the name doesn't refer to storage; it refers to the number of bits of data the processor could operate on at one time. And megabyte means not "1,000,000 bytes" but something slightly larger, as a peek under the hood will reveal.

It's commonly known that a bit (short for *binary digit*) represents a single indivisible piece of information. The only values it can represent are 0 and 1, which isn't enough to hold a full name or birthdate or Social Security number. But just as the lowly dot and dash can be combined to make all the letters, digits, and symbols of Morse code, combinations of 0s and 1s can be used to represent arbitrarily complex data. A pair of bits can be interpreted as any of four values (00, 01, 10, and 11), three bits can take on any of eight values (000, 001, 010, 011, 100, 101, 110, and 111), and so on.

With each additional bit, the range of values doubles. A group of eight bits can encode any of 256 values, which is enough to assign one unique value to each letter, digit, punctuation mark, or special symbol from a large number of Western languages. This unit is called a byte (or an *octet* in countries where the deliberately silly play on the word "bit" isn't as hilarious as it is in English*). A group of sixteen bits is sometimes called a word (playing off its two-byte length, which is enough to spell out short words like "ah" and "uh" and "ew" — an entire day's conversation for some programmers). When viewed as a single 16-bit value, a word can

---

*   Four bits, or half a byte, is called a nybble. Ten bits is a decle, five bits is a nyckle — yes, it's all very sylli.

represent any of more than 65,000 symbols, enough to cover nearly every writing system on Earth.

Digital devices use base two because it implements the purest form of either/or thinking (see "Nuh *Uh!*"); if all you know about a value is that it's not a zero, then it must be a one, and vice versa. This makes error correction and recovery of lost data more reliable than in any other base, including our familiar base ten. Since only two values are pre-approved, any rogue value that happens to appear (caused by, say, a groggy transistor) is taken as representing whichever of the two it more closely resembles. (Some would say this is also how two-party elections work.)

The memory in Sheldon's Nintendo 64, and in all computer systems and gaming consoles, doesn't just store letters and numbers and symbols. Every piece of information that the system must accept or manipulate or display — every sound effect or dot of color or bonus multiplier, every keypress or mouse movement or thumb twitch, along with the instructions for processing those pieces of information — must be converted to a stream of 0 and 1 bits and stored (at least temporarily) in the machine's memory. Sheldon's Nintendo needs to keep track of a lot of information, so it needs a lot of megabytes of storage.

Strangely enough a megabyte, whose name (according to "Atto Way!") sounds as though it ought to equal exactly 1,000,000 bytes, actually equals 1,048,576 bytes. Similarly, a kilobyte equals not 1,000 but 1,024 bytes. Because of engineering constraints, computer memory is most commonly available in sizes that are powers of two. It just so happens that 2^{10} (two to the tenth power) is pretty close to 1,000 ($2 \times 2 \times 2 \times 2 \times 2 \times 2 \times 2 \times 2 \times 2 \times 2 = 1{,}024$), and so rather than inventing a whole new prefix (what's the Danish for "1,024"?), computer memory designers simply shrugged and coopted the prefix kilo- to mean thousand-ish. In the same way, a megabyte (2^{20} bytes) happens to be just a little more than a million bytes (a million-ish), and a gigabyte (2^{30}) works out to a billion-ish (1,073,741,824).**

Given that a byte is eight bits, the sixty-four megabytes in a fully tricked-out Nintendo represent over half a billion distinct pieces of information: enough to give a 0 or a 1 to every person in North America. This seemed like a lot at the time — certainly to Nintendo's marketing department. After all, the first IBM PC XTs were capable of supporting only one percent of that amount: 640 kilobytes of memory. But that's still a lot when compared to the 128 kilobytes on the original Macintoshes seen in both Howard's and Leonard's bedrooms, which is twice the sixty-four kilobytes available in the Commodore 64, one of the leading personal computers up to that time, which dwarfs the sixteen kilobytes on the original IBM PC, released only as many years ago as actor Kunal Nayyar was.

** Since there are only about a million different values that can be represented by a combination of twenty bits, and since every question in the game of Twenty Questions can only be answered "yes" or "no," the claim has been made that there must be only about a million different things in the world. Just as sylli.

Think nothing could ever be done with so few bytes? The computer in the Command Module on the *Apollo 11* spacecraft had less than 40,000 words of memory: the equivalent of one Commodore 64 and one IBM PC. That's not enough memory to store one low-resolution snapshot from a smartphone, not enough for more than a handful of obnoxious ringtones, but more than enough to fly three guys to the Moon and back.

But we're not done yet. The PC's sixteen-kilobyte memory size dwarfs the five kilobytes found on the first computer to sell one million units, the VIC-20. Those first million VIC-20s combined had less memory than a single iPod Nano has.

Still, the VIC-20 was an improvement on the four kilobytes in Radio Shack's Tandy TRS-80, one of the first mass-produced personal computers. Four kilobytes is only enough memory to store half the text of this chapter. What could *you* do with a four-kilobyte computer?***

The microcomputer and game console industries were born in 1971 with the invention of the single-chip microprocessor: the Intel 4004. It operated with a separate memory chip that had a capacity of forty bytes. *Forty.* Not enough room even to say *HELLO*

*** If you were Apple Computer's Steve Wozniak, the answer to the question "What could you do with a four-kilobyte computer?" is "Sell several million of them." That was the amount of memory in the 1977 Apple][that Sheldon asks Woz to sign.[1] And yes, that's a pair of square brackets standing in for the Roman numeral II in the name of the Apple][. No less sylli than the three slashes in the name of the company's third-generation offering, the Apple ///.

WORLD DO U LIKE MY CAT PHOTO LOL ;) — or something equally vital.

All of these figures refer to RAM, which is erasable memory. The memory in the Nintendo's cartridge is almost entirely ROM (non-erasable), a convenient way for most gaming consoles of the era to distribute their games. Yet as visual textures, animations, and live-action sequences became more complex, video games outgrew cartridges. By the late 1990s, games were typically being distributed on CD-ROMs: compact discs containing software instead of audio tracks. CD-ROM games took much longer to start up than cartridge-based games did, but a single disc, holding 650 megabytes or more, could include as much data as ten Nintendo cartridges.

Unlike CD-ROMs and other non-erasable memory technologies, the rewritable memory in most computers requires a constant supply of electricity. If the machine loses power, it suffers instant amnesia. Personal computers get around this problem by using rewritable media, such as hard disks and thumb drives, to store snapshots of their state of mind, but many early models had no hard drive at all. One or two floppy disk drives were the most they could offer in the way of non-volatile removable storage. (Ever wonder why a PC's hard drive is called "C:" instead of "A:"? It's because the earliest PCs had only a single floppy drive (named "A:"), along with an empty slot for installing a second (named "B:") and no hard drive. If you wanted a second floppy drive, you had to shell out. If you wanted to install a hard drive (which then became "C:"), you *really* had to shell out.)

The ancient floppy disk Sheldon proudly displays, the one containing his list of mortal enemies, most likely holds just over one megabyte, but the earliest floppy disks had less than a tenth of that capacity.[2] Few end users of today can imagine thinking in such a small space. Modern laptop computers usually boast at least a few gigabytes of memory and several hundred gigabytes of hard drive space. The lowest-capacity DVD can hold as much as seven CDs: 4.7 gigabytes of data, or almost six bits for every person on the planet. (Enough for three shaves and haircuts apiece.)

And as internet connections grow faster and downloading becomes less time-consuming, users have been demanding ever-larger devices for storing all the junk they'll never have the time to look at. A $100 one-terabyte disk drive, about the size of a piece of toast, holds 2^{40} or a trillion (-ish) bytes, equivalent to a stack of bare CDs as tall as a person. And ten-terabyte drives will no doubt soon become commonplace . . . or will they?

Surely no one'll *ever* be able to use *that* much storage. Right?

ASK AN ICON: Gordon Moore

Gordon Moore (Caltech Ph.D. '54) is the cofounder of Intel Corporation and a life trustee of Caltech. He's well known for Moore's Law (given that name by Carver Mead), an observation on how many more components can be crammed into an integrated circuit each year relative to the year before.

In 1965, Moore published a paper analyzing the maximum number of lowest-cost components a single semiconductor chip could hold. He noted that in the six years since their invention, that number had doubled every year, a trend that he supposed would continue for at least another decade.

Engineering companies took Moore's assessment as a call to action, and it became something of a self-fulfilling prophecy. But certain physical limitations are inescapable, and the annual doubling couldn't continue forever, as Moore himself pointed out in a 1975 reassessment. He predicted that the rate would cool somewhat in the short term, with the doubling now taking two years rather than one.

Nevertheless, Moore's Law is continually being trundled out by marketers and pundits alike, most of whom distort and recast it into "X doubles every eighteen months," where X is whatever technological or economic indicator the speaker fancies (computing power, microprocessor performance per dollar, hard drive capacity, etc.).

Moore and his wife, Betty, have donated $200 million for the construction of the Thirty Meter Telescope (see "The

Naming of Things"), and their $600 million gift to Caltech in 2001 remains the largest single academic donation in world history.

Q: Sheldon refers to imagination as "the world's most powerful graphics chip."[3] As the power of graphics hardware continues to soar, do you foresee a time when silicon will outperform the human imagination?
Gordon Moore: Silicon is already outperforming the human imagination. Close your eyes and let your mind take you somewhere amazing. Anywhere — let your imagination run wild.

Now realize that no matter what scene your mind's eye has dreamed up, people have already seen something just like it, but even more breathtaking, more detailed, more vivid, on the screen of a movie theater, or a computer, or a video game console.

In stereo.

---

1. "The Cruciferous Vegetable Amplification" (Season 4, Episode 2)
2. "The Russian Rocket Reaction" (Season 5, Episode 5)
3. "The Irish Pub Formulation" (Season 4, Episode 6)

TWELVE
MY NUMBER'S BETTER THAN YOURS

> Sheldon: What is the *best* number? By the way, there's only one correct answer.
> Raj: 5,318,008?
> Sheldon: Wrong. The best number is 73. You're probably wondering why.
> Leonard / Howard / Raj: No. / Uh-uh. / We're good.
> — "The Alien Parasite Hypothesis" (Season 4, Episode 10)

It appears that the *Big Bang Theory* writers decided to honor the show's seventy-third episode, and actor Jim Parsons's birth year, with a little paean to the number 73. And what makes 73 the "best" number? According to Sheldon, it's because:

* 73 is the 21st prime number;
* the mirrors, or decimal reversals, of 73 and 21 are 37 and 12;
* 37 is the 12th prime number;

* multiplying the digits of 73 gives 21;
* and the binary representation of 73 is 1001001, a palindrome (that is, it reads the same backward and forward).

The first and third of these facts are certainly true (a prime number is any positive integer having exactly two positive integer factors), but the other three are merely cute curiosities. Yes, 73 is an emirp (a prime whose decimal reversal is also a prime), but that's noteworthy only because we happen to count in base ten. That accident of mathematics arose from the use of fingers as counting devices (hence the two meanings of the word "digit"), and by making it a cornerstone of his argument Sheldon displays an anthropocentric bias quite uncharacteristic of him. (That bias is absent from a promo filmed at about the same time, in which the characters break the fourth wall to discuss the fact that *The Big Bang Theory*, having entered syndication, will now air five nights a week. Sheldon proposes making the announcement in base five, "a numbering system that would dominate if human beings had a total of five fingers." This leads to a contemplation on the ramifications of "a human race with five one-fingered hands.")

As for his observation that 73 is a palindrome in base two, that's only mildly remarkable. With 0 and 1 being the only valid digits (see "Making Memories"), many numbers reveal themselves to be palindromes in that base — including 73's sidekick, 21 (10101), and its reversal, 12 (001100), though not 37 (100101).

As Leonard points out, an argument about which number is the "best" comes perilously close to being a good conversation-killer. On the other hand, some might say that an argument about which shoe designer is the best, or which rock band or which number of fingers per hand, could be just as meaningless — and, in the right company, just as much of a conversation-killer.

"It's difficult to work in a group when you're omnipotent." All right, that's a quote from *Star Trek: The Next Generation*, but it could easily apply to Sheldon's state of mind. In the absence of an actual diagnosis, and since we know he's not crazy (his mother had him tested), what is it with that guy's attitude?* Why would he presume to dictate which number is "the best"? The word *science* means "knowing," but what we "know" is a matter of opinion and is subject to change (see "Nuh *Uh!*"); does he not "know" this?

One of the reasons teamwork is often a better problem-solving strategy than working alone is that the diversity of input gives greater opportunities for insight. Yet Sheldon typically shuns collaboration, disparages the contributions of others, and fires those who presume to share credit.[1] With few exceptions, it seems that he only joins teams in order to exercise power, as with the company to create a handwriting recognition app, or because the rules require it, as with the Physics Bowl team.[2]

Laura Scherck Fizek, M.S.W., is a professor of organizational behavior at Suffolk University in Boston. She took a look at some aspects of Sheldon's behavior in groups: his condescension, his intolerance of interruptions, his failure to interpret social cues, his perpetually misguided belief that others are counting on him to do the intellectual heavy lifting. Then, using the Physics Bowl episode as a case study, she analyzed the effects of these traits on others.

Though Sheldon's bossiness strikes most viewers as a chief reason for the dysfunction or dissolution of virtually every team he joins, Fizek doesn't see that as the underlying cause. People aren't motivated to stay on teams unless their individual roles give them satisfaction, she says, and in the absence of a shared purpose it's impossible even to define those roles: "When you aren't in alignment around why you've formed the team, then you're not a team." In the case of the Physics Bowl there's no single common purpose, so there's no team to speak of. Leonard, Howard, and Raj have one objective throughout: to win the competition as a team. Sheldon has no interest in competing, let along collaborating, and agrees to team up with the others only because (according to his interpretation of Spock's dying words) their needs outweigh his. Once he's been kicked off the team, his motivation changes, and he now vows revenge, although to his mortification he finds himself unable to accomplish it singlehandedly. (Interestingly, the quiz question that stumps him appears to be the only one that isn't a matter of simply recalling facts.)

Don't blame Sheldon, Fizek cautions; he stays consistent. His premise never changes: he begins as an individual, and he ends as one. Moreover, she applauds him for the self-revealing way in which he introduces his teammates as "the third-floor janitor, the lady from the lunchroom, and — my Spanish is not good — either her son or her butcher."

In Sheldon's world, that little confession of uncertainty constitutes extraordinary humility.

Or it may just be a reminder that he doesn't much care about teams and teammates.

---

*   Sheldon certainly shows many of the symptoms associated with the autism spectrum, most notably with Asperger's syndrome, but the show's creators have made it clear that he does *not* have any definite diagnosis. That may change once the next edition of the *Diagnostic and Statistical Manual of Mental Disorders* comes out, but in the meantime it's probably sufficient to file him under "Incredibly Obnoxious Personality Not Otherwise Specified."

Calculator Fun

Raj's vote for 5,318,008 as best number, as he later explains, is because when you enter 5318008 into a calculator and look at it upside-down, it spells BOOBIES. (Sheldon misses a rebuttal opportunity here: 73 upside-down spells EL — Superman's family name.)

Calculator spelling works best on inexpensive calculators featuring seven-segment displays, in which each digit is constructed entirely out of straight lines (the kind often seen on sports scoreboards and clock radios). Viewed upside-down on such a display, the digits 0 through 9 resemble the block letters O, I (or l), Z, E, h, S, g (or q),

L, B, and G (or b). This somewhat restricted alphabet can nevertheless be used to spell out quite a few words, many of them well aligned with the refined tastes of cheap-calculator-owning youngsters with too much free time. But once those same individuals have grown up to become astrophysicists and neurobiologists, they tend to outgrow their seven-segment displays and buy themselves graphing calculators, a far more powerful breed. (Mayim Bialik's jazzy TI graphing calculator can display graceful curves and intricate multi-line equations in beautiful typefaces while keeping her from inadvertently typing naughty words — except perhaps for $\int e^{x-y!}$.)

Actress Jessica Walter, who plays science patron Mrs. Latham,[3] also appears on the TV comedy *Arrested Development*, wearing a prison uniform with the number 07734 emblazoned across the front. This means that every time she glances down, her, er, 5,318,008 say hɛLLO.

In what sense can the "goodness" of a number even be measured? Some numbers are just funky. Some just have funky names. An example is the class of numbers known as taxicab numbers. Early in the twentieth century, the mathematician Srinivasa Ramanujan (after whom one of the characters on the CBS crime drama *NUMB3RS* was named) lay ill in a London hospital. His good friend the mathematician G.H. Hardy (after whom the Hardy Boys were *not* named) came to visit him. After they'd shared a few root-beer-in-a-square-glass jokes and caught each other up on the latest installment of the Taylor series,

Hardy mentioned that he had noticed the number of the taxicab that had brought him. (Of course he did. Did you know he was a mathematician?) It was #1729, which had struck him as a rather dull and perhaps inauspicious number. (Proof that any particular number can be more or less exciting or auspicious in the first place is left as an exercise for the reader.)

Dull, G.H.? Hardly!

Hardy never explained what he meant by "dull." It may have slipped his mind that 1729 was the year of the founding of Baltimore — or perhaps it didn't — but in any case he had evidently been unaware of some of 1,729's more intriguing properties, such as that in decimal it's the product of the sum of its digits and that sum's decimal reversal:

$$1 + 7 + 2 + 9 = 19$$
$$19 \times 91 = 1,729$$

Doesn't get much more exciting than that, eh?

Ramanujan, however, got quite excited (he was a mathematician too, after all), remarking that 1,729 is actually a very interesting number, for the reason that it is the smallest number that can be expressed as the sum of two cubes in two different ways. That is, if $x^3 = x \times x \times x$, there's a pair of numbers a and b that will satisfy $a^3 + b^3 = 1,729$, where $0 < a \le b$, and there's a second pair that will also satisfy it. (Can you figure out all four numbers? If you can't be bothered or can't find the time [betcha you could find the time if you were in the hospital with only visits from mathematicians to look forward to], there's a hint at the end of the chapter.)

Other numbers can also be expressed as the sum of two cubes in two different ways (such as $4{,}104 = 2^3 + 16^3 = 9^3 + 15^3$), but 1,729 happens to be the smallest. That makes it rather special, and that's what had caught Ramanujan's eye.

In honor of this story, and because inside every mathematician beats the heart of a poet, an entire class of numbers has been named the "taxicab numbers." (We could lay out the precise definition here and say that the nth taxicab number is defined as the smallest number that can be expressed as the sum of two cubes in n different ways, but we recognize that when it comes to making dull things interesting, we're no Ramanujan.) The first taxicab number is 2, since it can be expressed as the sum of two cubes in one way ($2 = 1^3 + 1^3$). Hardy's cab number, 1,729, is the second. It's also known as the Hardy-Ramanujan number, and perhaps you can guess why. (Mathematicians don't, however, call it that in daily usage. When little mathematicians are playing tag, they don't say, "1,727, 1,728, Hardy-Ramanujan number, 1,730 — ready or not, here I come!" That would just be weird. But ask any mathematician, "What's a taxicab number?" and you will instantly hear this little tale about Ramanujan and Hardy and their non-dull taxicab, whether you want to or not.)

Should we get excited that 1,729 has this property? Does it serve any useful purpose to those of us who aren't mathematicians or taxicab spotters? Perhaps not (or not yet, at any rate), but you never know. Many numbers with similarly oddball properties have turned out to be useful in unanticipated ways. This very number did just that for Ramanujan, who had discovered this property years before and had merely been waiting for a chance to drop it into a conversation. (That doesn't ruin the story for you, does it?) One day, someone may turn the taxicab numbers into the basis of an advanced cryptography system or a better method of packing objects into rigid containers (mathematicians into taxicabs, perhaps?) or something even more unanticipatable.

Anyway, that's just an example of a number that some people might argue has more of a right to be called the "best" number than boring old 73.

EUREKA! @ CALTECH.EDU
"Go BEAVERS!"
Rumors to the contrary notwithstanding, Caltech does not award athletic scholarships to its incoming freshmen. There's no need. Undefeated (or, more accurately, unopposed) on the gridiron since 1993, the school's football squad until recently held the record for most games played at the Rose Bowl (none of them on New Year's Day, alas) and once shut out every one of their opponents over an entire season (namely 1944, when many of the country's finest athletes were otherwise engaged). In baseball and men's

basketball, Caltech boasts the longest continuous streaks in NCAA and conference history (though not necessarily *winning* streaks).

Dubious statistics aside, it is in fact the case that Caltech athletes have ranked nationally in table tennis, fencing, and Ultimate. This comes as a surprise to many, thanks to society's cruel stereotypes about the love of numbers and the love of sports, as demonstrated by the difficulties 3D chess and blindfolded Rubik's cube–solving continue to face in becoming Olympic events. And yet, some of the biggest nerds around are found among the fans of professional sports. The eagerness of these walking databases to spout statistics and predictions unbidden puts Sheldon's "One Thousand and One Digits of Pi" kindergarten recitation to shame.[4]

Caltech political scientist Rod Kiewiet teamed up with former Los Angeles Dodgers general manager Fred Claire to teach "The Theory and Practice of Moneyball," a seminar on the analysis of objective baseball information. Sabermetrics (named after the Society for American Baseball Research) is the science of mining ballpark statistics for useful or informative patterns. With nearly every baseball team now employing a professional sabermetrician, some of the correlations that have been uncovered are astonishing. It turns out, for instance, that batting average is a poor predictor of wins and losses.

As techniques for collecting raw data continue to improve, sabermetrics has developed into its own legitimate sub-branch of economics. It's subject to many of the same laws, constraints, and strategies that govern money.

Not to mention that with its help, your fantasy baseball team might finally make you some serious real-world dough.

---

* USC has the Trojans, Notre Dame has the Fighting Irish, Caltech has "our friend the beaver."[5] (Yes, thanks, we get it. What can we say — when the mascot was being voted on, giant tube worms hadn't been discovered yet.) And would you believe that one of the official songs of the Caltech football program was "Lead Us On Our Fighting Beaver"? No doubt about it: "in a world where mankind is ruled by a giant intelligent beaver" (to quote Sheldon and Amy's game of "Counterfactuals"[6]), "Nature's engineer" probably finds it pretty hard to be taken very seriously.

OUT TO LANDS BEYOND

"In order to be effective and credible, I needed to understand the science."

Tom Lloyd (Caltech Ph.D. '99) calls himself "a poster child for weird career paths." At breakfast on day one of his first year in college, he realized that what he wanted to do most of all was to be a rock-and-roller.

A classically trained cellist, Lloyd had intended to major in music and philosophy. Instead he cofounded a band, and as soon as he finished his freshman year, he dropped out of college to tour with them. Featuring Lloyd on bass and Dan Zanes on lead vocals and crazy hair, the Del Fuegos performed throughout the world for ten years. Their hits included "Move With Me Sister," "Don't Run Wild," and "I Still Want You." (That's three separate song titles, incidentally — or one *very* fascinating telegram.)

After the band members finally dis-membered the band, Lloyd, now 28 years old, was determined to pick up his studies where he had left off, although with a change of focus. He majored in bioresource sciences at Berkeley before earning his doctorate in environmental engineering science at Caltech. Later, when his interests shifted again, he became a management consultant, then a research analyst, and he's now vice president of a global investment management firm.

And for your information, his thesis was on the mechanism by which an iron-binding compound called deferrioxamine B dissolves minerals containing transition metal oxides. There's a "rock" pun in there somewhere, but it would take someone with a Ph.D. to make it.

How to Find Two Pairs of Cubes Whose Sum Is 1,729

Make a list of numbers and their cubes:

$$1^3 = 1 \times 1 \times 1 = 1$$
$$2^3 = 2 \times 2 \times 2 = 8$$
$$3^3 = \ldots$$

You can stop once you've found a number whose cube is bigger than 1,729.

Now find a pair of cubes from your list whose sum is 1,729. Then find a second pair.

Look — you're a mathematician!

---

1. "The Cooper-Nowitzki Theorem" (Season 2, Episode 6)
2. "The Bus Pants Utilization" (Season 4, Episode 12) and "The Bat Jar Conjecture" (Season 1, Episode 13), respectively.
3. "The Benefactor Factor" (Season 4, Episode 15)
4. "The Vacation Solution" (Season 5, Episode 16)
5. "The Justice League Recombination" (Season 4, Episode 11)
6. "The Zazzy Substitution" (Season 4, Episode 3)

THIRTEEN
KISS AND TELL

> Leonard: Sheldon, why is this letter in the trash?
> Sheldon: Well, there's always the possibility that a trash can spontaneously formed around the letter, but Occam's razor would suggest that someone threw it out.
> — "The Cooper-Hofstadter Polarization" (Season I, Episode 9)

The snide comment comes and goes without explanation, and neither Occam nor any of his tonsorial accessories are mentioned again. But whether or not you've heard the expression before, you can probably guess at Sheldon's meaning just from the context: something along the lines of "Go with the most likely explanation." The principle he's referring to is sometimes expressed as: "Entities must not be multiplied beyond necessity." This is often reduced to "Don't overthink" or "KISS" ("Keep it simple, stupid!"), but it's a little subtler than that. Occam's razor is a rule of thumb, not a law. It's a reminder that whenever we're trying to understand the connection between two entities or events, we should start by examining the simplest plausible explanations.

Observation teaches us that most of the time, things that seem unrelated *are* unrelated. Say you're trying to estimate what tomorrow's high temperature will be. You might take into account such factors as today's weather, the average temperature this time of year, and conditions elsewhere on the planet, because they would all seem to have some bearing on your estimate. But you probably wouldn't take into account things like how much time you'll spend on your daily walk, which party has a majority in the Senate, or whether a woodchuck could chuck wood, because it's unlikely tomorrow's high will depend on any of those things. And if it turns out it does, you'd surely want to know why.

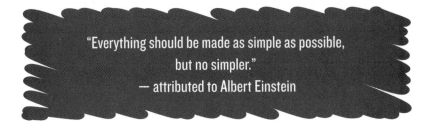

"Everything should be made as simple as possible, but no simpler."
— attributed to Albert Einstein

It takes time to whittle down a hypothesis to just the relevant cause and effect, but that whittling is vitally important. A cause is deemed both *necessary* and *sufficient* when it's enough to produce a given effect — and just enough. If you can get away with less, then some part of the proposed cause isn't necessary;

if you need more, then it's not sufficient. Aside from giving a more accurate picture of Truth, removing all the irrelevancies tends to give hypotheses a certain elegance. Want to pay a scientist a huge compliment? Just say, "My, what an elegant hypothesis!"

Irrelevancies aren't always symmetrical. The chill in the air won't depend on how briskly you walk as much as how briskly you walk depends on the chill in the air. Your speed varies inversely with the temperature. (Then again, you're warm-blooded; as Sheldon speculates, if lizards had weathermen, they might say things like "Bring a sweater — it's slow outside."[1])

Since the laws that govern our Universe tend to mind their own business, the simplest description of reality usually comes closest to the truth, whether we like it or not. (And we often don't like it. In the original pilot for *The Big Bang Theory,* which has never been aired, a proto-Penny character named Katie listens unhappily while Leonard recites a list of poor life choices she's made. When she finally demands to know whether he's saying her stupid choices mean she's stupid, he mumbles awkwardly, "Well . . . the simplest equation is usually the right one."[2])

Sometimes what appears to be a complication turns out to be a simplification. You'd think that since an apple falls to the ground and the Moon doesn't, there must be two different laws at work: one for things that fall to the ground, like apples, and one for things that don't, like moons:

* Why does an apple fall? Because the Earth pulls it downward.
* Why doesn't the Moon fall? Because the Earth, uh, *doesn't* pull it downward?

But that's not right. Sir Isaac Newton applied Occam's razor and realized that an apple falls down and the Moon doesn't, not for different reasons but for the *same* reason. In both cases, gravity is the cause. The difference is that the apple is relatively motionless, while the Moon is moving sideways to the Earth at high speed — zooming past us, as it were. Without gravity to draw it toward us, continually bending its straight-line path into a curve, it would continue on an unwavering course out into space (see "'Round and 'Round"). Newton had discovered the law of universal gravitation — and not, as is commonly believed, gravity itself (as if until he came along apples hadn't fallen or people hadn't noticed).

Why had the answer been eluding everyone for so long? Because they had been asking the wrong question. Instead of asking, "Since the Moon is up above the Earth, why doesn't it fall down like an apple?" they should have been asking, "Since the Moon is in motion, why doesn't it go screaming off into the void like the kid at the end of the line in crack-the-whip?"

* Why does an apple fall? Because the Earth pulls it downward.
* Why doesn't the Moon shoot off into space? Because the Earth pulls it downward.

Same reason, same law.

Many occurrences that seem to be independent from one another actually aren't, and what looks like a complicated interaction turns out to be a simple one. You might have marveled at the curious way the Moon always shows the same face to us as it goes around the Earth. In the time it takes the Moon to rotate once on its axis — a little less than a month — it revolves once around the Earth in the same direction. It's like a circling sumo wrestler who never turns his back.

You don't have to go looking for a complicated explanation, because those two physical phenomena are closely related; in fact, the one is the cause of the other. Billions of years ago, the Moon's speed of rotation wasn't the same as its speed of revolution. Whoever — or whatever — was prowling around at night back then could look up and see its different faces at different times.

But the mass of the Moon isn't evenly distributed; it has a heavier face and a lighter one. If you toss a rubber duck into a bathtub, it will bob and roll on the water until gravity finally brings it to rest with its weighted bottom pointing (usually) straight down or (rarely) straight up. The same thing happened to the Moon, and it now floats around the Earth with the same face always pointing "down" (that is, toward us).

On the other hand, some quantities that seem to be intimately connected are actually completely unrelated. The Moon and the Sun, as seen from Earth, are nearly the same size, so much so that during a solar eclipse the Moon can almost exactly cover the Sun. You might think there's a significant reason behind that, but it's just a coincidence of diameters and distances. The Sun just happens to be 400 times wider than the Moon and 400 times farther away from us.

In its youth, when the Moon was closer to Earth than it is now, it easily blotted out the Sun. As it continued to recede it diminished in size, and billions of years from now the Sun will appear much bigger than the Moon. Sometime after that, the Sun will expand drastically and will look bigger still. (Eventually it will swallow the Moon and the Earth, and that will be the end of that.) We just happen to be living at a particular time when the Sun and the Moon, as seen from Earth, appear to be about the same size. Just a coincidence. Honest.

It takes some deftness to wield Occam's razor. For one thing, it's not always clear which of two possible descriptions of reality has fewer elements. Which is more likely: that everything you see and hear and feel really exists, or that you're just a brain floating in a vat of chemicals and being fed by a supercomputer that makes you "think" you see and hear and feel it all? (While it would be pretty tricky to build a computer like that, it would be easier than creating an entire universe, wouldn't it?) Or try this one: is DNA something that people make in order to produce more people, or are people something DNA makes in order to produce more DNA? (Our egocentric view of the world tells us it's the former, but from an evolutionary standpoint the latter makes more sense.)

And sometimes the simplest possible explanation isn't the correct one at all. Evolution has left a lot of redundant information in our DNA. If you were designing humans from scratch, you might be able to come up with a much more elegant way to encode their genetic profiles. But your simple and elegant design wouldn't match reality.

Occam knew this. (Not the parts about super-computers and DNA.) He wasn't saying, "The simplest explanation is always the correct one." He was saying, "Do yourselves a favor and start your analysis with the simplest explanations. Don't drag in more complex explanations until you absolutely need to. Take a tip from Thoreau: Simplify, simplify." (Not that Occam ever read Thoreau.)

A simple explanation should be simple all the way through, not just on the surface. There's nothing preventing a random assortment of matter from spontaneously forming itself into a trash can (or a woman, as in Penny's surprisingly erudite physicist joke),[3] but the odds are strongly against it. It's only slightly more likely that five seconds from now all the air molecules surrounding you will suddenly decide to rush ten feet to the left. A physical description of either event would be fairly straightforward, but any explanation to account for that description most likely would not ("Uh, an alien molecule-manipulating ray?"), and any explanation that demands an inelegant hypothesis is suspect.

A (probably apocryphal) story about avoiding unneeded complexities concerns the French mathematician Pierre-Simon, marquis de Laplace. Chided by Napoleon for having published a complex treatise on the mathematical laws governing the Universe but never once mentioning its Creator,

> **Who of Where Said What?**
>
> *Occam* refers to a fourteenth-century English philosopher named William, who was evidently from Ockham, a town near London. *Razor* is a metaphor for the fine line that distinguishes the necessary from the needlessly complex.
>
> As for Occam's razor: (1) it predates William of Ockham; (2) he never expressed it the way it's most often quoted; and (3) when he did say it, he was referring specifically to questions of theology, not science.
>
> That's fame for you.

Laplace is supposed to have replied stiffly, "Sire, I had no need of that hypothesis."

No doubt the story, which can still raise hackles today, has been grossly embellished. If Laplace made any comment on the subject of the supernatural, it was probably only to refute a speculation by Newton that the celestial mechanism requires occasional adjustment by Someone (not a Creator — more a sort of celestial Tune-up Master). The larger debate — on whether postulating a deity simplifies or complicates things — has raged throughout history. It lies at the heart of Sheldon's struggle with his mother's worldview:

> Mary: Here in Texas, we pray before we
> eat. . . . [They pray.] Now, that wasn't so hard,
> was it?
> Sheldon: My objection was based on
> considerations other than difficulty.[4]

Not that science has all the answers. At present, we're aware of a fistful of physical constants that seem to be unconnected to one another — at least, so far they've resisted all attempts at simplification. Light travels at a particular speed in a vacuum, and no one knows why. Electrons have a particular charge and a particular mass, and no one knows why, but their values don't seem to be connected to the speed of light in a vacuum. The gravitational field around a given mass has a particular strength, and no one knows why, but it doesn't seem to be connected to the charge or mass of electrons or the speed of light in a vacuum. And so on.

About twenty of these fundamental constants have been identified. Each one has some particular value that it just happens to have, and no one knows why. Nor has anyone found a way to derive any of the constants from any of the others or to convert one to another. The expectation is that all of them *are* somehow interdependent and that eventually we'll figure out how. But at the moment, they're as unrelated to one another as are Tuesday, purple, and Alaska.

In the meantime, we can treat them individually, which keeps the physical laws that apply to each of them simple. But it also means we have a big collection of unrelated physical laws — one for the speed of light, one for the charge on the electron, one for its mass, and so forth — making our picture of the Universe more complex.

Do we *need* to distill all the laws down into one super-law? Nature is already shoving us in that direction by revealing some surprising exceptions to Occam's rule of thumb. The Atomic Age has shown us a connection between mass and all sorts of quantities that we might prefer to think of as distinct from it: energy (in Einstein's famous formula), time (in general relativity), and temperature (in black holes). Then there's the quantum gravity problem. For analyzing the behavior of super-small things that aren't very massive, the laws of quantum mechanics work just fine; for analyzing super-massive things that aren't very small, the laws of gravitation are perfect. But when you get into the realm of things that are super-dense (both super-small and super-massive), then both sets of laws come up short. Black holes are an example of something that can be super-dense. Another is the Big Bang theory itself (the one with the lowercase T, not the TV show): the notion that everything was crowded together, nothing was happening, and then suddenly ("nearly fourteen billion years ago") it all went boom and the Universe was created.

If it's true that the Big Bang was the first thing that ever happened in the Universe, then wouldn't whatever caused it have happened first? But then what caused *that?* Science is largely a search to understand the causes of things, yet the Big Bang appears to be an effect without a cause. For some people, this version of Genesis is the paragon of simplicity; for others, it's the antithesis. How attractive to imagine that something as vast and complex as our Universe rests on a foundation of pure nothingness, and how equally attractive to imagine that it doesn't. Both sides can claim to have Occam in their court.

With or without Occam's razor, some causes, such as the origins of life, have proven only slightly easier to unearth. Our reluctance to doubt the evidence of our own imperfect eyes accounts for the longevity of naïve beliefs like spontaneous generation, the notion that life arises from non-living material.* That idea predates Aristotle and wasn't discredited until the 1850s, largely due to the work of Louis Pasteur. It's an understandable mistake. Nothing seems deader than a chunk of dung, until living dung beetles emerge magically from it. How can something so un-alive suddenly turn into such a cornucopia of life?**

We now know that the reason dung beetles emerge from balls of dung is that their mommies laid their eggs there when no one was looking. A dung beetle egg is a pretty tiny speck, and we can forgive the ancients for not having noticed that they had come out of living beetles — and vice versa. While we're at it, let's forgive them for not knowing what it is that gives people hay fever or causes plague or makes fuzzy

---

*   Spontaneous generation is not to be confused with regeneration, which is the ability of certain creatures to regrow various tissues after losing them. Not hair and feathers, which grow from beneath the skin, but more complex structures, such as an iguana's tail, an axolotl's leg, or an earthworm's entire rear(ward) end.

**   Their apparent ability to arise from nothing is what made the dung beetle, or scarab, so sacred to the ancient Egyptians. It wasn't the dung per se that excited them but the magic of life. As for Penny's pal Zack, who believes that "if you kill a starfish, it'll just come back to life" — well, he didn't learn *that* from the ancient Egyptians.[5]

stuff grow on old food. And now that we're feeling comfortably patronizing toward the ancients, let's pause to reflect that in some sense they were right about that spontaneous generation stuff. Once upon a time, somehow, somewhere in the Universe, life *did* arise from non-living material.

Put *that* on your razor and slice it, Occam!

Marcolli's cross-disciplinary perspective makes for an unusually powerful tool. For instance, to address the gaps in our understanding of the Big Bang, she'll propose exotic and largely speculative scenarios of the early Universe and then work through the math to see which sets of hypothetical starting conditions could result in the Universe we see today. Or she'll recognize that a certain physical process neatly demonstrates what was previously considered a purely mathematical concept. In such cases, a deeper examination of the physics can sometimes provide greater insights into the math.

Her work may help inch us closer to a longstanding dream of physics: a "theory of everything" (TOE) — a description of reality that would unite all known physical laws into one elegant super-law. That's a tall order, one that Einstein worked on for years before he ran out of time. Under a TOE, all the physical constants that currently seem to be completely unrelated to one another (the speed of light, the strengths of gravity and electromagnetism, the mass of the electron, and so forth) would be understood as interconnected manifestations of a single underlying cause, just as an apple and the Moon are no longer thought of as obeying separate laws of motion.

One single, elegant law to cover life, the Universe . . . and everything. One Rule to ring them all. William of Ockham would be so pleased.

That (Other) Historic Apple

Later in the "Occam's razor" episode, Sheldon tries to make a rhetorical point about scientific inspiration by asking scornfully, "Was the apple falling on Newton's head . . . just an anecdote?"[6]

Gee, Sheldon, yes, it was. There's no evidence that Newton was ever hit on the head — other than metaphorically — by an apple.

Why isn't Sheldon aware of this? His behavior at this moment is more than a little irrational, so perhaps his acute excitability is causing him to be temporarily misinformed on the history of science. But Leonard, still in control of his own faculties, responds in a way that suggests that he also believes the apple-on-the-head story.

One of Newton's biographers did once quote him using the example of a falling apple when discussing gravity, and another said that seeing a particular apple fall is what had given Newton his initial insight, but neither of them reported any sort of actual fruit beaning.

There's a lovely, if rather gnarled, apple tree still growing on Newton's family estate in England, said to be the one from which The Apple fell. To this day, apples still fall from it (though of course not far from it). It could easily have inspired Newton without directly assaulting him.

Surely anyone smart enough to invent calculus could figure out how gravity works without intervention from Granny Smith.

1. "The Recombination Hypothesis" (Season 5, Episode 13)
2. 2006 pilot (unaired)
3. "The Dead Hooker Juxtaposition" (Season 2, Episode 19)
4. "The Electric Can Opener Fluctuation" (Season 3, Episode 1)
5. "The Justice League Recombination" (Season 4, Episode 11)
6. "The Cooper-Hofstadter Polarization" (Season 1, Episode 9)

FOURTEEN
ONE POTATO, TWO POTATO

> Arthur (Professor Proton): I power a clock with a potato.
> Penny: . . . Wouldn't that solve the world's energy crisis?
> — "The Proton Resurgence" (Season 6, Episode 22)

If only. Sadly, the lowly potato doesn't put out enough juice per unit weight to make it a viable solution to the global energy problem. You'd need several thousand tons of them just to power a small electric car (not pounds: *tons*), and the fleet of trailers it'd take to haul them around would put a serious dent in your mileage — to say nothing of your parking options. The problem isn't that "potato electricity" is somehow different from "battery electricity" or "wall outlet electricity" or "lightning electricity" or "rubbing a balloon in your hair" electricity — it's not. It's just that a common tater can't store or deliver enough of it.

All electricity boils down to the position and movement of electric charges. As for what electric charges are made of, we don't know, other than that they come in two flavors that behave as exact opposites of each other. Benjamin Franklin arbitrarily named these flavors "positive" and "negative," and we continue to call them that because their behavior can be modeled using the arithmetic of positive and negative numbers, and because we don't have anything better to call them.

In electricity, as in life (and certainly as in *The Big Bang Theory*), opposites attract. A positively charged object and a negatively charged object will feel a mutual attraction. Whenever anything intervenes to move them physically farther apart, this force acts to push them closer together. Objects whose electrical charges have the same sign (both positive or both negative) will feel a mutual repulsion. Whenever anything intervenes to move them physically closer together, the force acts to push them farther apart. The strength of this bidirectional force falls off in proportion to $1/r^2$, the square of the distance between the objects, the same way gravity does (see "The Gravity Situation").

All protons, it turns out, have a charge of +1, and all electrons have a charge of –1. Within a molecule each particle's position matters, but at much larger distances the effects overlap as if all the charges had simply been added together. An object's net charge, whether atom or aircraft carrier, is the sum of all its particles' charges. So an atom with as many electrons as protons (the default state) has no net charge, while an atom with unequal numbers of electrons and protons (called an ion) always has a non-zero net charge.

The net charge on most everyday objects stays close to zero, because an excess of either type of charge typically creates a flow of free electrons (electrons that aren't bound to any atom at the moment) between the object and its environment, resolving the imbalance. A dramatic demonstration of this effect is seen in so-called heat lightning, which occurs when strong winds within a thundercloud split electrically neutral water droplets into flurries of negatively and positively charged ions. The extra electrons in the negative flurries are repelled by one another and are attracted toward the positive flurries. As the flurries grow, more ions accumulate, until at last the drama reaches its striking climax.

free electron An electron that — at least temporarily — isn't married to an atom.

Electrons are the gadabouts of the subatomic world; unlike atoms and ions, they can move easily through many materials, jumping from atom to neighboring atom. Electrically powered devices work by capturing energy from the motion of electric charges, typically free electrons scurrying through metals and semi-conductors. And where do those free electrons come from? Often from generators or batteries, devices for separating charged particles. Generators rely on mechanical means; batteries rely on chemical means.

generator A device that separates charged particles mechanically, such as by using a moving magnet to pull them along a wire or by depositing them onto a moving belt of material and scraping them off elsewhere.

battery A device that separates charged particles chemically, such as by ionizing some of its own atoms and physically moving the resulting ions and free electrons.

At each of the two terminals of a battery, a chemical (the electrolyte) reacts with a piece of material (the electrode) in such a way as to separate positive charges from negative ones, drawing one off more than the other and leaving the electrode with a net charge. (For example, an acid electrolyte might erode a metal electrode little by little, stripping entire atoms out of it while leaving some of their electrons behind. The positively charged metal ions float away in the acid, and the surplus of electrons gives the electrode, or what's left of it, a negative charge.) So that the charges on the two electrodes will be unequal, a different chemical process is used at each terminal. (The + and – markings near a battery's terminals aren't necessarily the signs of their net charges; they only mean "relative to each other.")

Suppose you're a free electron and a chemical or mechanical process has placed you at the negative terminal of a battery or generator. You're surrounded by other free electrons and perhaps some negative ions, all repelling you with their

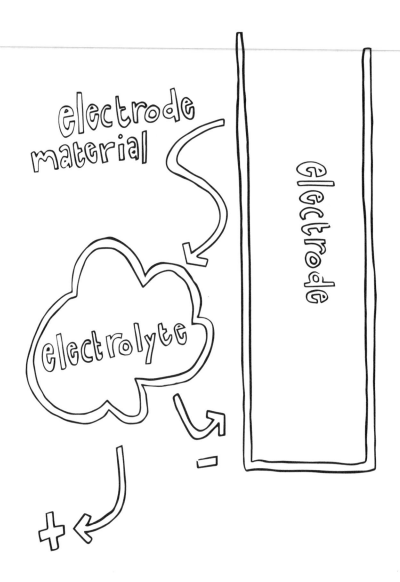

negativity. Meanwhile, over at the positive terminal a gaggle of positively charged ions are all beckoning to you while simultaneously repelling one another. You'd like nothing better than to race right across the battery or generator and get up close and personal with a lovely positive ion. But alas, the same processes that created and separated the two of you in the first place are keeping you apart. What to do, O subatomic Romeo?

Who Ate My Foil?
When lasagna (or a similar dish) is kept in a steel baking pan and covered closely with aluminum foil, the result may be an inadvertent "lasagna cell." Pan and foil form the electrodes of this (not very efficient) battery, and acidic tomato sauce is the electrolyte. As the chemical reaction proceeds, it may erode holes in the foil and deposit black spots of oxidized aluminum onto the food.
O yum.[*]

Provided the two terminals are connected electrically (say, by the two wires of a load such as a motor or a lightbulb or a stereo), you can go the long way around, following the path of (literal) least resistance. You exit the battery or generator via the wire, pass through the guts of the device (giving up a little energy of motion as you go), and return by way of the other wire to the positive terminal, where your beloved(s) await. In fact, you don't have to run all that way yourself; an electron in the far wire will gladly jump into the positive terminal, making room

***

[*] Surgeon General's Warning: Lasagna, potatoes, or any other food being used as a battery gives up its right to be used as food.

for some electron behind it to move into its place, allowing another one to move into that one's place, and so on all the way back to your neck of the woods, where an electron at the very tip of the wire will move slightly deeper into it, making room for you to jump into its place. The energy that makes all this happen is a combination of the push from your negative neighbors and the pull from your positive admirers. And that's how an electric current flows.

A single potato can't produce enough electrical push and pull to turn the big, bulky motor of an old-school analog clock, but it can handle the much lower power requirements of a small digital clock or wristwatch (the kind with an LCD or LED readout). Electric power is a combination of current and voltage: current measures how many electrons are passing through the innards of the device each second, and voltage measures how much energy is being used to push each of those electrons through. A digital watch requires much less energy than a clock motor, so it can get by on a power source that delivers fewer electrons per second (current) at a lower energy per electron (voltage) — such as a potato.

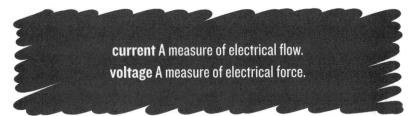

current A measure of electrical flow.
voltage A measure of electrical force.

The maximum voltage or current a battery can deliver depends on the speed and strength of the chemical processes by which charges get separated out and sequestered at the two terminals. As long as no load is attached to the battery, an ion or free electron has very little incentive to crowd into a confined space already full of other ions and free electrons, all of which are repelling it. Beyond a certain limit this resistance outweighs the abilities of the chemical process, and no additional charges can crowd in. In the case of an acid that strips positive ions out of a piece of metal, the electrode's negative net charge increases until it's attracting ions back as fast as they're being extracted.

Once a load is attached to the battery and electrons begin flowing from the negative terminal to the positive terminal, the net charge on both terminals moves closer to zero, making room for new electrons and ions to be sequestered there. If the battery fails to replenish those charges as fast as they're being drained off, then the voltage drops, the current drops, the battery drops (to its metaphorical knees), and everything grinds to a halt. It's possible to trade current for voltage, as when a car's ignition coil converts the strong current coming from the low-voltage battery to a weak current at a much higher voltage. But a flashlight battery produces a low current at a low voltage, which is why a handful of flashlight batteries is no replacement for a car battery. Their chemical processes can't deliver enough high-energy electrons per second to meet the engine's huge electrical demands. The electrical trickle from a potato is even weaker; it can reach almost as high a voltage as an ailing flashlight battery (typically a volt or so), but with less than a percent of a percent as much current. The poor spud is just not that studly.

How can we get more juice out of our tuber? Let's dig a little deeper.

A single electrolyte-electrode combination is called a half-cell. Putting two half-cells together so that ions and/or electrons can flow between their electrolytes makes a two-electrode structure called a cell, and a battery consists of one or more cells. For example, a twelve-volt car battery consists of six two-volt cells, while a nine-volt household battery (your smoke detector could do with a fresh one, by the way) contains six tiny cylindrical 1.5-volt cells in two rows of three.

half-cell A single electrolyte and electrode.
cell Two half-cells with their electrolytes connected.
battery One or more cells connected electrically.

The two half-cells that make up a cell can contain two different electrode materials, two different electrolytes, or both, as long as the electrodes settle at different voltages. In the case of the potato battery, two different electrode materials are bathed in a shared electrolyte: the potato's own naturally occurring phosphoric acid. But it's a weak acid, and not much metal is exposed to it, so it doesn't produce a great deal of voltage or current. What's needed is . . . more potatoes.

A Batch of Battery Batter

Battery technology is centuries old, but the unexplored frontiers of chemistry are so vast that new electrode and electrolyte combinations continue to appear. Recent breakthroughs in the trade-off between battery weight and power have made electric cars a common sight. Much of this development work is being alternately driven and stalled by automobile manufacturers (Raj's sister Priya even helps set up "a secondary derivative market which would allow overseas car firms to hedge their investments against potential advancements in battery technology"[1]).

Common electrolytes include the alkaline paste in flashlight batteries, the sulfuric acid in car batteries, and the lithium salt in some cell phone batteries. There's even a battery that runs on human urine. What's most important is for the electrolyte(s) to create positively and negatively charged products by reacting chemically with the electrodes, and for the electrodes to reach different voltages. (Yes, urine.)

For this reason, if a cell contains only one electrolyte (as is the case with the potato battery and similar devices, such as the lemon battery and the sauerkraut battery), then the two electrodes must be made of different materials. Any two dissimilar metals will usually work, such as a piece of wire and a nail, or a nickel and a paperclip, or two recent (copper-zinc) pennies, one with some of its copper scraped off.

(Yes, urine. Can we please move on?)

How the cells in a battery are hooked together makes a difference. Connecting them in series (with the negative terminal of each connected to the positive terminal of the next) gives a higher voltage than a single cell but the same maximum current. Connecting them in parallel (negative to negative and positive to positive) gives the same voltage as a single cell but provides a greater maximum current. Combining these wiring schemes gives the best of both worlds. We can increase the voltage by connecting a few potatoes in series; then we can increase the current by making up several hundred such sets and connecting them in parallel. Or we could do it the other way around: connect several hundred potatoes in parallel, then make a few such sets and wire them together in series.

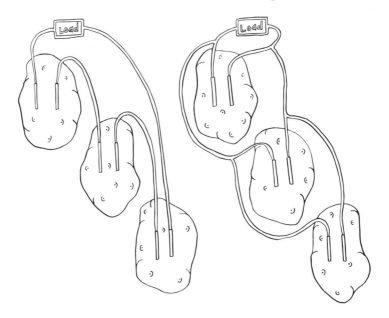

Three potatoes connected in two different ways. **At left:** Connecting them in series triples the voltage. **At right:** Connecting them in parallel triples the maximum current.

And now at last we can light a flashlight bulb or turn a small motor or blast some awful music.

Incidentally, the potato is far from the only kind of do-it-yourself battery around, though perhaps it would like for us to think it is.** It's really only acting as a stiff framework for propping up two metal electrodes in a chemical bath. We could get the same result with, say, an orange or a stale jelly doughnut. Or a sponge soaked in vinegar or antacid or oven cleaner.

Or, yes, urine.

** Then again, who can say whether a potato thinks, or, if so, what it thinks about (see "Coming to Think of It")? On the other hand, it's obvious what a sweet potato thinks about: "I think, therefore I yam."

EUREKA! @ CALTECH.EDU
Fuel Pardon the Pun

Fuel cells are another type of solid-state device for delivering electrical energy. Like batteries, they work by temporarily separating electrons from their atoms so that in their mad rush to recombine, they can be made to do useful work. But while a battery's "fuel" is entirely self-contained, in the form of bits of solid material being chemically eroded away, the fuel for a fuel cell must be pumped in from the outside.

As atoms of fuel enter the cell, they pass through a material that strips electrons off them. The resulting positively charged ions flow into and through a material that's impervious to free electrons; consequently, the electrons must take the long way around, through an electrical circuit. At the far side of the material, the emerging ions encounter the returning electrons in the presence of another substance from the outside, typically atmospheric oxygen. All three ingredients combine, and the resulting molecules are discharged from the cell as waste products.

Caltech's Sossina Haile researches solid-state ionic materials, with an emphasis on their use in thermochemical fuel production. She invented a specific type of solid-acid fuel cell in which the ionized fuel undergoes a process called the superprotonic phase transition.* As with most fuel cells, her cells can run on something as simple and plentiful as hydrogen gas, in which case what comes out of the cell's discharge tube is pure water. To prove this, Haile once drank the runoff from a fuel cell. Try *that* with your car's battery! (Actually — please don't.)

Concepts for fuel cells have been around for almost two centuries, but it's only recently that improved manufacturing techniques and lower costs have made them an attractive energy-production alternative. And why not? Hydrogen is cheap, easily managed, and plentiful; in fact, it's one of the two most abundant elements in the Universe. (The other, according to writer Harlan Ellison and musician Frank Zappa, is stupidity.)

---

*   Superprotonic — what a great name for a high-tech startup, right? Already taken . . . by two of Haile's grad students.

Out to Lands Beyond

"I feel a certain kindredness of soul with anyone who squandered a technical degree . . . to go do art."

Robert Lang (Caltech '82, Ph.D. '86) majored in electrical engineering and applied physics and went on to author or co-author over eighty scientific papers. But he couldn't shake his childhood passion: origami, the ancient Japanese art of paper folding.

With his engineer's and artist's eye, he developed software to compute folding patterns for producing three-dimensional objects of arbitrary complexity. At the same time, he began devising origami-based solutions to problems in engineering. He now works full-time as both an origami artist and an industry consultant on foldable structures, from super-compact automobile airbags to a space telescope lens that fits into a rocket but unfolds to the size of a soccer field.

The author of several books on origami, Lang is credited with over 500 original designs. These include dinosaurs, scorpions, and a Black Forest cuckoo clock ("It works! It tells the correct time twice a day!"), each of them folded — as is de rigueur in origami — from a single sheet of paper, with no gluing, no cutting, and no tearing.

---

1. "The Cohabitation Formulation" (Season 4, Episode 16)

FIFTEEN
SMOOTS AND WHEATONS

> Sheldon: I'm looking for a barber and I'm running out of time. My hair is growing at the rate of 4.6 yoctometers per femtosecond. I mean, if you're quiet you can hear it.
> — "The Werewolf Transformation" (Season 5, Episode 18)

It sounds as though Sheldon's pulling terminology out of thin air (if not thin hair), but yoctometer and femtosecond are real words (see "Atto Way!"), and they really do denote a (very short) length and a (very brief) timespan, respectively. Converting them to something a little more familiar, like meters and seconds, is only a matter of shifting decimal points.

But that's because the metric system is based on powers of ten, which are easily manipulated in our decimal system. By contrast, the people who designed the customary system used in the U.S. for distances, areas, weights, volumes, and so forth seem to have had anything but ten fingers: twelve inches to the foot, 640 acres to the square mile, sixteen ounces to the pound, nine teaspoons to the jigger, etc. For the entire rest of the world, this continues to provide both endless amusement and endless irritation — an ancient and honorable American tradition.

Practically the only non-metric unit of measure that corresponds to a power of ten nowadays (besides the penny, the dime, and the tenths-of-a-mile digit on your odometer) is the mil, which represents one 1,000th of an inch. A nineteenth-century holdover, the mil is used mainly to report the thickness of items such as plastic sheeting, trash bags, and other twentieth-century holdovers.

One notable attempt at decimalization predates the mil by more than two centuries: Gunter's chain, a sort of measuring tape used for marking off distances along the ground. This sixty-six-foot lightweight metal chain consists of 100 metal links about eight inches long, with a brass ring marking every tenth link to speed counting.

Workin' on the Chain, Gang

The surveyor's chain invented in 1620 by Edmund Gunter elegantly converts between rods (1 rod = 5^1/$_2$ yards) and acres (1 acre = 4,840 square yards) using only decimals. By design, a square measuring one chain by one chain (four rods by four rods) has an area of exactly one tenth of an acre, so to find the acreage of any rectangular plot of land, one simply multiplies the length (in chains) by the width and divides by ten. For example, a piece of ground having a length of two chains plus forty links, and a width of one chain plus twenty-five links, covers 0.3 acres (= 2.40 × 1.25 ÷ 10). A twenty-chain-by-twenty-chain square would enclose the proverbial forty acres. (Mule sold separately.)

This arithmetic is much easier than measuring length and width in (say) yards and having to divide the product by 4,840. It works because Mr. Gunter intentionally chose the length of his links to be a power-of-ten fraction of the square root of a power-of-ten fraction of an acre. Clever!

And no, he wasn't American.

Almost anything can be turned into a unit of measurement. One night in 1958, at an institute of technology somewhere in Massachusetts, diminutive freshman Oliver Smoot's fraternity brothers pressed him into service as a living yardstick for measuring the length of the Harvard Bridge across the Charles River. The final tally: 364.4 smoots, plus or minus an ear.[*]

Another unit of measure based on a contemporary last name, made famous by the comic strip *Dork Tower*, is the unofficial unit called the wheaton, which honors actor and frequent *Big Bang Theory* guest star Wil Wheaton. By definition, a one-wheaton Twitter account has 500,000 followers.[1] The unit was

---

[*]   Smoot is the cousin of Nobel physicist George Smoot, who appears as himself in one episode of *The Big Bang Theory*[2] and whose last name Penny finds so amusing (pretty rich, coming from someone who's gone for years without even *having* a last name), unaware perhaps of its currency (at least in some circles).

defined when Mr. Wheaton, an avid social media user, first became able to make that claim for himself; his account is now several wheatons in size. Most people's Twitter followings are more conveniently measured in milliwheatons or (presumably even-numbered) microwheatons.

Some perfectly legitimate units of measure have names that misleadingly suggest that they're outright jokes (such as the barn, the shed, the outhouse, and the skilodge, which are all measures of area on a sub-atomic scale). Other names sound imprecise: shouldn't the meaning of one horsepower or one man-hour depend on which particular horse or man is providing the labor? At the North Pole, Sheldon notes that medical help is "eighteen hours away by dogsled."[3] Though he doesn't actually utter the word "dogsled-hour," it would be perfectly acceptable to propose such a unit, provided he could get agreement on suitable standards for weather, terrain, sled design, snow slickness, canine attitude, etc. Like the light-year (the distance light travels in a year), the dogsled-hour would be a measure not of time but of distance.

In that same episode, Leonard defines a standard hug as "two Mississippis, tops," a notion reinforced in a later episode by a would-be suitor of Penny's.[4] They're referring to a popular technique for maintaining a (not very accurate) one-beat-per-second counting tempo by reciting the word "Mississippi" in between counts. Other people say "one thousand" before (or after) each count, apparently in the belief that jumping up and down the number line doesn't in any way make counting confusing.

If it's exotic units Sheldon is after, then instead of yoctometers per femtosecond, he could have chosen the furlongs per fortnight favored by engineers. This deliberately humorous quantity, incorporating two legitimate but archaic units of measure (a furlong is an eighth of a mile and a fortnight is two weeks), works out to almost exactly one centimeter per minute, or about the strolling speed of your average starfish.

Furlongs per fortnight aren't the only measure of speed that happens to convert fairly cleanly to different units. One mile per hour is just about one and a half feet per second (because 5,280 feet/mile divided by 3,600 seconds/hour is very close to 1.5). The speed of light (a little under 200,000 miles per second) is close to a foot per nanosecond.

A yoctometer per femtosecond (10^{-24} meters per 10^{-15} seconds) is the same as 10^{-9} meters per second. This is comparable to the unofficial unit of length known as the beard-second (the length the average beard grows in one second). Without any loss of precision, Sheldon could have declared his hair to be growing at a rate of 4.6 nanometers per second.**Although that wouldn't have made him sound any less geeky or annoying, he would at least have been done talking that much sooner.

Some terms are only meaningful in a hand-wavy, approximate way. There's no standardized

---

** If he were willing to sacrifice a *little* precision, he could have said "almost half a millimeter a day" (4.6 nanometers per second times 86,400 seconds per day). For reference, half a millimeter is about ten times the diameter of a single hair.

definition of a car length (as Sheldon points out to a DMV employee[5]) or a city block, but the images they convey are generally sufficient for the purpose. A hairsbreadth ranges from 10 to 100 microns, or about 1% to 8% the thickness of a dime. Alternatively, having a precise definition doesn't guarantee usefulness. When we use a football field to indicate length, are we supposed to remember not to include the end zones and the regions beyond them?[***] The dog year is an explicit unit of time (one-seventh of a year), but it's as meaningless as the notion of linearly accelerated canine maturation it's based on. For that matter, there's no scientific basis to the notion of audible hair growth, either. But try telling Sheldon that. He'll claim he can't hear you over the noise of his hair.

[***] We'd be better off using a football field to indicate area, since the playing area, from one five-yard line to the other, covers just about one acre. (For Canadians, measure from one team's seventeen-and-three-quarters yard line to the other.)

EUREKA! @ CALTECH.EDU
Femto-Slo-Mo
Caltech's Ahmed Zewail, the "father of femtochemistry," won the Nobel Prize for his method of freeze-framing chemical reactions. The system works by launching a reaction and then taking a very high-speed laser "snapshot" a few femtoseconds later, while the reaction is still in progress. Though only one snapshot can be taken each time the reaction is triggered, the

operation can be repeated many times while varying the time delay between trigger and snapshot.

The resulting collection of snapshots, each taken at a different point in the reaction, can then be interleaved to create a "movie" at the equivalent of several trillion frames per second.

To view an entire second's worth at normal video frame rates would take tens of thousands of years. Fortunately, the reactions Zewail studies are over in the blink of an ion.

In What Universe?

It's Off to Work We Go

If we can't nail down the exact position of the apartment, can we at least restrict it to one region of the city? A possible way to attack this problem is by tracking the route Leonard takes to commute to his job at Caltech. Indeed, the route is the subject of one entire episode, where we learn that work is far enough away that driving there takes more than a few minutes, bus service is an option, and a Pottery Barn may be along the way.[6] We're also told that Leonard prefers Los Robles Avenue over speed bump–infested Euclid Avenue, just one block over.

The speed bumps are a writers' fabrication, but those street names are real. Then again, so is Michigan Avenue, mentioned by Sheldon in passing, and since it's much closer to Caltech than Euclid, it would make for a more sensible approach route.* (For that matter, so is Catalina Avenue, which, though not mentioned in the episode, features prominently in its working title, "The Catalina Alternative." In the handful of days between table read and filming, "Catalina" was replaced with "Euclid," evidently as homage to the ancient Greek mathematician of that name — or perhaps just because it sounds funnier.)

Things get even stranger when Penny drives Sheldon to work. Though they arrive at their destination, they evidently haven't reached their journey's end. At one point on the commute, recognizable landmarks of Caltech's famously compact campus are seen flashing by the car windows. Minutes later, she's still driving him. The same mysterious occurrence happens in another episode. "Hey, this isn't the way to work," Raj cautions Howard — just as campus buildings begin to flash by.[7]

Perhaps there's an astrophysical explanation. In the same way that time appears to an outsider to grind to a halt at the event horizon of a black hole, it may be that it takes an infinite amount of time to pass through the Caltech Event Horizon. But if that's the case, then once in, how does one ever escape?

Many graduate students have asked themselves that same question.

---

* We're assuming that Leonard prefers a sensible commute. At least once, we see him cruising to work in the 800 block of South Raymond Avenue, half a mile the wrong side of Los Robles from Caltech.[8]

1. http://dorktower.com/files/2009/05/DorkTower721.gif
   shortcut: http://DaveZobel.com/-bbdt
2. "The Terminator Decoupling" (Season 2, Episode 17)
3. "The Monopolar Expedition" (Season 2, Episode 23)
4. "The Misinterpretation Agitation" (Season 8, Episode 7)
5. "The Euclid Alternative" (Season 2, Episode 5)
6. Ibid.
7. "The Locomotion Interruption" (Season 8, Episode 1)
8. "The Vacation Solution" (Season 5, Episode 16)

SIXTEEN
ABOVE THE BELT

The trans-Neptunian object Raj has discovered is just a small player on a huge stage. It isn't even a real planet, any more than Pluto is (see "The Naming of Things"); it's only a planetary object, meaning an object with planet-like geology. But it's impressive enough to get him a mention in *People*, a magazine with twice the circulation of *Popular Science* (presumably because people are twice as popular as science).

trans-Neptunian object (TNO) An object whose orbit around the Sun takes it beyond the outermost planet (Neptune, three billion miles out).

Sheldon has said that he'd only be impressed if Raj's TNO plummeted toward Earth and Raj "exploded it with his mind." And that's good thinking, since we should be concerned any time an object's trajectory crosses Earth's orbit. The tiniest ones make for pretty meteor showers, but anything larger than about a golf ball can do serious damage: shock waves, tsunamis, dust clouds, even extinctions. (Not to mention landing on people, which appears to have happened only two or three times in all of recorded history.) After all, it was most likely a lump smaller than Washington, DC, that collided with the Earth and wiped out the dinosaurs sixty-five million years ago. And thousands of buildings in Russia were wrecked in February 2013 by the shock wave coming off the mid-air explosion of a meteor no larger than a house.

But since Raj's object is "beyond the Kuiper belt," that puts it at least five billion miles away (for the moment, anyway). Not much of a threat. The objects we need to be watching out for are much, much closer.

> **Kuiper belt** Rhyming with "wiper," it's a two-billion-mile-wide ring around the Solar System, beyond Neptune's orbit. It teems with chunks of material, many big enough to qualify as dwarf planets. The largest denizen of the Kuiper belt is Pluto. (Pluto's nemesis, Eris, inhabits the scattered disc, a region that extends even farther out.)

We're accustomed to thinking of space as practically empty, but that doesn't mean collisions can't happen. (Very little of the airspace above a rifle range contains actual bullets, but it's still not the safest place to loiter about in.) The Solar System is crammed with orbiting chunks of rock and ice, including many that cross Earth's path. We're plowing through a perpetual cosmic hailstorm.

The biggest threat to us comes from the near-Earth asteroids, which number in the millions. We know they're out there, and a single fairly small one could easily take us down, *Deep Impact*– or *Armageddon*-style (or at least interfere with a surgical procedure, as Sheldon unhelpfully reminds Leonard).[1] But we've never spotted most of them, let alone computed their orbits. If Raj really wants to be popular with the people (people who read *People*), he might consider pointing his telescope a little closer to home.

That's Okay, Leonard,
No One Says "Hofstadter" Right Either*

The peculiar name given to Raj's discovery has the form of the provisional designations assigned to such objects by the Minor Planet Center of the International Astronomical Union. Leonard's pronunciation ("two zero zero eight en cue sub seventeen") is rather labored; he could have just gone with "two thousand eight en cue seventeen." But either way is still faster than "the 441st of all objects that were first seen between July 1st and 15th, 2008," which is what that alphabet soup translates to.**

This was a deliberate television fabrication: in actuality, during that specific timeframe fewer than 150 new objects were observed, but that was an exceptionally quiet two-week period. Since 1999, the MPC has assigned, on average, more than 100 new provisional designations *a day*.

---

* Sometimes we hear Leonard's last name pronounced with a short A sound, sometimes with a long one, sometimes with an E. We've also heard Leslie Winkle's first name said with an S (instead of the Z she herself gives it). And almost no one but Amy hits the hard J in Raj.

** The 2008 is obvious, the N means the thirteenth half-month (A is 1 and I is omitted), the Q means "the 16th object discovered" (omitting I again), and 17 means "+ (17×25)." That is, 441 = 16 + (17×25).

Ground-based telescopes like the ones Raj uses are useful for tracking all that space debris, but they can see only half the sky at any given time. The other half is below the horizon. A telescope in orbit around the Earth (like the Hubble Space Telescope) or around the Sun (like the James Webb Space Telescope now under construction) can see much more of the sky.

The B612 Foundation, named after the asteroid on which Antoine de Saint-Exupéry's Little Prince lived (and not after the *Batman #612* Stuart sells to Wil Wheaton,[2] in case you were wondering), plans to place an infrared space telescope in orbit around the Sun. Completing one circuit every seven months, it will continuously search for and track near-Earth asteroids. Any found to be on a collision course with us will be handled not by exploding them, Sheldon (do we really want to turn one big ball on a known trajectory into many little ones spraying out randomly?), but by using the gravitational pull of a "space tugboat" to shift their orbits ever so slightly.

EUREKA! @ CALTECH.EDU
Manna(rals) from Heaven

Caltech mineralogist Chi Ma is one of the few people who might actually be pleased to see rocks falling from the sky. His job is to crack meteorites open and train an electron microscope on their innards. Many of the objects he studies are fragments of asteroids dating back almost five billion years, siblings of the giant boulders that once crashed together to form our own home planet.

And what has he found? More than twenty minerals never before seen — some of which had never even been imagined by scientists.

If those materials ever existed here on Earth, they were churned away long ago by continental drift, erosion, and the general environmental unpleasantness that characterized the planet's infancy. Only those left out in the cold, empty vastness of space were preserved. Now, freed from their geologic time capsules, they give us a glimpse of some of the first solids formed in the just-born Solar System.

Definitely old school.

ASK AN ICON: Rusty Schweickart
Rusty Schweickart, pilot of the *Apollo 9* lunar module, is cofounder and chair emeritus of the B612 Foundation (b612foundation.org). The organization is spearheading the hunt for objects on collision courses with Earth. (Howard is in illustrious company: the school near Harvard where he earned his master's degree also granted one to Schweickart.)

Schweickart's classic essay "No Frames, No Boundaries" is a poignant meditation on Spaceship Earth from the perspective of a voyager traversing the heavens — which, in a sense, is all of us.

Q: How can we avoid getting wiped out by space junk like the dinosaurs were?
Rusty Schweickart: Finding trans-Neptunian objects is great

science, and happily it can be done using ground-based telescopes, just as Raj did. But the near-Earth asteroids actually pose a threat to the survival of life on Earth, and they're much smaller and darker than TNOs or comets.

While ground-based telescopes have helped us find the ten thousand largest NEAs, there must be at least a million smaller ones that we haven't found so far. Sentinel, B612's specialized space telescope, is designed as the "early warning" system to do this job.

It's the largest environmental project of all time.

---

1. "The Septum Deviation" (Season 8, Episode 9)
2. "The Russian Rocket Reaction" (Season 5, Episode 5)

SEVENTEEN
SAYS YOU!

Sheldon: Gossip? I'm disappointed in you.
Amy: Now, now. Evolutionary biologist Robin Dunbar has identified gossip as an aid to social bonding in large groups.
Sheldon: Forgive my language, but — poppycock.
— "The Herb Garden Germination" (Season 4, Episode 20)

Er, that's an aid, not a requirement, Amy — but let's let that go for now and try to find something good to say about gossip. Who knows? Maybe we'll end up feeling better about the way we treat our friends when they're not around.

Robin Dunbar is a British professor of evolutionary psychology, and according to him language is "the single most important evolutionary development in our history as a species."[1] Whether or not you agree, it's natural to assume that language was created as a way of sharing important information about our prehistoric surroundings. Surprisingly, however, that may not have been the case at all.

Though it's hard to know what early man had to talk about that was so vital, it's reasonable to suppose that he relied on his newly acquired facility for speech at least as much for social purposes as for technological reasons. (How long could it possibly take to cover all the nuances of "we hunt now," "you drop rock," "me smell panther"?) Dunbar found a similar balance in modern-day speech patterns: he discovered that we devote as much as 60% of our conversation time to subjects involving relationships and personal experiences rather than the physical world.*

After all, we're primates, and primates are highly social animals. They form social groups and engage in lengthy mutual grooming rituals for the express purpose of maintaining frequent personal contact. Once early humans had tamed their environment (and themselves) and social groups had begun growing unprecedentedly large, maintaining those many bonds became a full-time task. It's possible that language evolved as a more efficient way to learn and share behavioral characteristics across the entire membership: long-distance grooming, as it were.

Gossip is the term Dunbar uses to refer to

---

* Interestingly, men — at least in younger age groups — spend about twice as much time discussing themselves as they do discussing other people; for women, the ratio is reversed. Dunbar sees an evolutionary imperative in this, tied to the critical areas of reproduction and child-rearing, with females seeking one another's support and advice, while the males focus on . . . advertising.

this socially directed use of language. The word's unpleasant connotations might make it seem ill-chosen, but Dunbar doesn't use it in a pejorative sense. What's commonly thought of as gossip — namely, the reporting of negative things about others with or without justification (dishing dirt) — actually takes up only around 5% of normal conversation time. The rest of the time, we're just yakking.**

What gives legs to some pieces of gossip and not to others is still largely unknown. But that doesn't justify the experiment Amy subsequently proposes to Sheldon: to track the flow of gossip by spreading a fairly tame but entirely mendacious rumor among their small social group. She must know that she can't reliably estimate, control, or prevent the psychological damage any individual or relationship might suffer, either from hearing the lie or from realizing later that it was a lie. As with all human experimentation, great caution is required (see "I Am the Very Model of a Modern Model Organism").

Amy's circle of acquaintances is probably smaller than most people's, but in any case primate social groups don't grow arbitrarily large. Within any particular group, the list of who's in and who's out fluctuates over time. In fact, an individual can be a

---

** Sheldon makes a (probably unintentional) play on words
   when he says, "Forgive my language." Whether or not
   he's aware of the two meanings of *gossip*, the thing that
   obviously displeases him the most isn't that Amy would share
   unflattering details of a friend's behavior, but that she would
   waste his time and hers discussing human behavior at all.

member of more than one group at once (a family, a hunting party, etc.), but no group ever has more than a certain number of members. The reason may be a simple matter of neurobiology. It turns out that there's a proportion between the maximum number of others with whom an individual consistently interacts at one time and the volume of the neocortex (the outermost — "thinking" — layer of the brain). The neocortex, Dunbar speculates, is crucial for maintaining relationships, and too many at once would simply overload it.

The only primate species that doesn't show any limit on group size is man. In fact, we don't know whether humans have a need for such a limit. Certainly, modern humans can and do interact (at least on some level) with potentially limitless numbers of individuals. Is this because our use of language makes our neocortex a more efficient relationship-processing engine? Or are we simply addicted to joining groups, regardless of whether we can manage them? If so, then based on the volume of the human neocortex, it could be argued that social order can be expected to break down once a personal group exceeds about 150 people (plus or minus fifty or so). This value, known as Dunbar's number, gives the maximum number of people one can theoretically keep in good social contact with (though Sheldon groans that he finds maintaining a mere five friendships to be "a Herculean task"[2]).

Dunbar's number isn't about keeping score of mere acquaintanceships — it's about stable interpersonal relationships. Evidence suggests that over the past quarter million years or so, the population of human social groups stayed considerably smaller than 150. But that's no longer the case. Our overcrowded, hyperconnected, postmodern world often forces us to be in social contact with many more people than that at one time, and our poor neocortices should be simply overwhelmed.

Science-based syndicated advice columnist Amy Alkon, author of *I See Rude People* and *Good Manners for Nice People Who Sometimes Say F*ck*, believes that this imbalance is at the root of the astonishing rudeness so prevalent in modern society. According to Alkon, since we didn't evolve to spend much time around strangers, we're not deterred from behaving badly toward them by the built-in self-restraint that's at work when we're with people we know. (Alkon lives in Los Angeles, three times voted the rudest city in America by the readers of *Travel + Leisure* magazine. New York has demanded a recount.)

Chuck Lorre's vanity card #210, seen at the very end of the first season's penultimate episode, acknowledges that mutual support has become harder to come by now that we live in towns and cities rather than in small tribes.[3] And this seems to be an unavoidable phenomenon, one that writer David Wong refers to as the Monkeysphere.[4] The telemarketer who interrupts your dinner, the driver who cuts you off, the troll who posts a criticism on your blog — all of them can directly impact your happiness and well-being, but since they exist outside your own personal Monkeysphere, *their* happiness and well-being mean almost nothing to you. They can't. We humans just don't have the brainspace for it.

What about virtual relationships: those happening only in the online world, which is (obviously) an animal unto itself? Without all the focused attention, time-consuming interruptions, and greeting and departing rituals required by real-world interaction, Dunbar's number may be too conservative. On the other hand, the quality of virtual relationships suffers from a lack of tonal cues, visual feedback, and instant accountability, and that breeds misunderstandings, rudeness, and cruelty. In any case, it's likely that our poor prehistoric neocortices can only handle so many interactions at once, whether in cyberspace or in meatspace — which is probably the reason your cousin's old college roommate unfriended you on Facebook last week.

Well, shucks, you barely remembered her name anyhow.

EUREKA! @ CALTECH.EDU

Aren't You Just a Little Curie-ous?

Without social bonding, schools would never work; without schools, society would never work. Institutions of higher learning have a vested interest in the primary- and secondary-school programs that feed them. It's a chain that supports itself.

That explains Caltech's heavy commitment to educational outreach, with a particular emphasis on science, technology, engineering, and mathematics (STEM) education. The Caltech Center for Teaching, Learning, &

Outreach coordinates partnerships with local educational foundations and school districts. The center also takes an active role in educating educators and offers courses in online education, flipped classrooms, and other emerging learning technologies. Director Cassandra Volpe Horii, whose doctorate is in atmospheric chemistry, is an expert in learner-centered teaching and curriculum innovation.

Caltech's student houses stay open year-round to support summer programs for high-school students and underrepresented groups. Pasadena's middle-schoolers are invited onto campus for the annual robotics challenge (see "Ro-, Ro-, Ro- Your 'Bot") and other special events. And the Caltech Precollege Science Initiative (CAPSI), a collaboration to create science and engineering programs for use in secondary schools, brings together some of the most talented educators and industry professionals on the planet.

Meanwhile, an army of volunteers recruited from the student body supports programs throughout the Los Angeles area. The Caltech Classroom Connection (CCC) sends student assistants into local K–12 classrooms on a regular basis. The volunteers accumulate teaching experience, while the youngsters get to enjoy some amazing learning opportunities, because as Amy declares, "[T]eaching is a performance art. In the classroom paradigm, the teacher has the responsibility to communicate, as well as entertain and engage."[5]

The Caltech Y, a student-life organization, provides volunteer service throughout the community. As one example, volunteers from the Y visited local non-profit Trash for Teaching (see "About the Author"), where they spent the day converting clean factory-discarded materials into kits for teaching science.

[SCIENCE TO COME]

Electrons Ain't Planets!

An Emmy-like ball of whizzing electrons caps Leonard's Physics Bowl trophy, Zack mistakes an atom on the cover of *Science* magazine for a planet, and if Sheldon could see the *Big Bang Theory* logo or the animated bumpers between scenes, he would be horrified. Little shiny balls spinning around bigger shiny balls on shiny circular paths? That's exactly what the electrons in an atom *don't* look like.[6]

It's not just the shininess or the visible paths or the wildly out-of-scale proportions. It's that the notion of atoms as "little solar systems" is terribly inaccurate. In some ways it gives us a useful mental image, but it's sloppy. An electron doesn't zoom around its atom like a little planet, with a definite position and velocity at every moment. In fact, whenever something isn't directly interacting with it, it's meaningless to say that it's "at" any specific location or is moving "in" any specific direction.

When you were an infant, experience taught you about object permanence: things continued to exist, with their properties largely unchanged, regardless of whether you were still observing them. With elementary (indivisible) particles like the electron, it's a little different. An electron in an atom inhabits a fuzzy, smeared-out "cloud" extending all around the nucleus. This cloud isn't a cloud of electrons. The reason it's a cloud isn't because something in it is moving too fast to see, like the blades of a fan. It's a cloud of something that we call "probability" (because we don't know what else to call it), something that takes the form of an electron only when anything's interacting with it. The rest of the time, it does nobody-knows-what.

Generally, when you probe the region of the cloud, you'll detect an electron. It will be somewhere(ish) and heading somewhere(ish) (see "It's a Fuzzy Old World"), but you can't predict where you'll find it and where it'll be heading — only their probabilities. And what you detect tells you nothing about what you'll find the next time you probe. In the in-between times, when you're not looking, not only don't you know anything about the electron, but there *is* no electron. There's only something that has the ability to conjure one up for you the next time you interact with it.

Unlike an electron, a photon is never confined to a cloud around an atom, but it leads a similarly fuzzy double life. How it behaves and where it appears depend on whether (and how) something else is interacting with it. In fact, all the elementary particles exhibit this duality. They aren't little hard pebbles; they're fuzzy somethings that behave like waves when we allow them to propagate as waves and like particles when we treat them as particles (such as by forcing them to cop to specific physical locations). And whenever we're not actively probing them, they inhabit

> a different realm of existence altogether (see "Slits and Stones").
>
> Yes, it's weird. And why shouldn't it be?

1. Robin I.M. Dunbar, "Groups, gossip, and the evolution of language," in *New Aspects of Human Ethology*, ed. A. Schmitt, K. Atzwanger, K. Grammer, and K. Schafer (New York: Plenum Press, 1997), 77.
2. "The Friendship Algorithm" (Season 2, Episode 13)
3. "The Peanut Reaction" (Season 1, Episode 16)
4. David Wong, "What is the Monkeysphere?" *Cracked*, September 30, 2007. http://cracked.com/article_14990_what-monkeysphere.html
    shortcut: http://DaveZobel.com/-bbms
5. "The Thespian Catalyst" (Season 4, Episode 14)
6. "The Bat Jar Conjecture" (Season 1, Episode 13), "The Justice League Recombination" (Season 4, Episode 11), and every episode ever, respectively.

EIGHTEEN
A THOUSAND POINTS OF LIGHT

Leonard: I'm not going to work.
Sheldon: Just because your career's been stagnant for a few years, there's no reason to give up.
Leonard: Sheldon, I was up all night using the new free-electron laser for my X-ray diffraction experiment.
— "The Euclid Alternative" (Season 2, Episode 5)

As it happens, what it is that's been keeping Leonard up all night isn't very important to the plot of this episode. The bottom line is that he's too tired to give Sheldon a ride to work — and from everyone's point of view, that's a crisis. Still, what *is* this X-ray diffraction experiment of his? (For the answer to the other obvious question — "what's a free-electron laser?" — see "Laser Fair.")

You can learn a lot about the microscopic structure of materials by seeing what they do when you shine a light on them. Even though you can't easily discern the individual molecules and atoms, there are shortcuts that can give you clues. One widely used technique is to get all the atoms of the material to arrange themselves in an orderly pattern. Their behavior in the aggregate then becomes easy to observe. The same principle is at work when you observe farm fields from a distance. Although you might not be able to make out the individual plants, the overall effect tells you at a glance whether you're looking at a cornfield, a rose garden, or a pumpkin patch.

A crystal is a repeating lattice of identical units in a geometric orientation. Familiar crystals include quartz, sugar, and frost on a window. Changing the structure can produce a different crystal: diamonds, graphite, and graphene are all made up of carbon atoms but in different arrangements.

crystal A piece of material consisting of a repeating pattern of atomic subunits, all identically constructed and oriented. Think of tiles on a floor, bricks in a wall, or cannonballs in a stack.

Disco balls are not atoms, and they don't generally form themselves into crystals, but they can help show what the atoms in crystals do. When a solo spotlight shines on the tiny mirrors covering a disco ball, half the room is filled with a thousand tiny points of light. The pattern of light depends strictly on the arrangement of the mirrors and the direction of the incoming light. In principle, given a blueprint of the disco ball, you could predict what pattern of dots a given beam of light will produce. This works the other way, too: if you couldn't see the ball, you could deduce the arrangement of its mirrors by varying the angle of the spotlight and observing the resulting light patterns.

This only works with a single ball and a single light, however. Add another ball or two, and it's instant chaos. Imagine trying to figure out which reflected dot came from which light striking which mirror on which ball. A crystal is like a cluster of disco balls, all constructed with the identical pattern of mirrors, all oriented in exactly the same direction, all being illuminated by identical spotlights in perfect alignment. Whether you have a cluster of two balls, or ten, or a thousand, each one would produce the same reflection pattern on the walls.

Unfortunately, thanks to the spacing between the balls, every pattern would be slightly offset from its neighbors. Except for the few blobs of light that just happened to overlap, you'd see one spot from every single mirror. That's a lot of spots to sort out. But a surprising thing happens if we move our cluster of identically constructed, identically oriented, identically illuminated disco balls into a much larger room. Imagine that the cluster is very far away from the wall — so far away that a person near the wall can scarcely distinguish among the individual balls. Now the light reflecting off any mirror on any ball will travel almost side by side with the light reflecting off the corresponding mirror on all the other balls, and the spots they produce will just about overlap. The original single-ball pattern emerges, and we can do our deducing-the-mirror-arrangement trick again.

Scientists don't spend a lot of time at discos, it appears (possibly because they'd spend all their time staring up at the spots on the walls), but they do use this cluster-of-disco-balls principle when studying crystals and crystal-like materials. By directing X-rays into a crystal and observing the patterns that form, they can deduce the construction and orientation of the crystal's subunits. As with the disco balls, each reflected spot consists of contributions from many subunits of the crystal.

There are two important distinctions to be made here. One is that while a mirror on a disco ball produces a blob of reflected light, the components of a crystal are so tiny that incoming X-rays aren't simply reflected. They're diffracted, spreading out in a wide curve like sound waves shouldering their way into a house through an open window (see "Slits and Stones").

The second distinction comes from the fact that the radiation in the X-ray beam, unlike the light from the spotlights, is deliberately collimated to resist spreading. As with the disco ball cluster, the crystal's components are so densely packed and so precisely aligned that X-rays diffracting off neighboring atoms travel almost side by side and arrive at the same place almost simultaneously. When they do, the slight differences in the lengths of the paths they've taken cause their crests and troughs to overlap in ways that sometimes produce a bright spot on the detector, sometimes a dim one. It's the same principle that's responsible for the light and dark bands of the interference pattern in the double-slit experiment (except that it's a zillion-slit experiment, with each component of the crystal acting as its own slit). Knowledge of the underlying physics then allows scientists to work backward from that visible pattern to the invisible structure that produced it.

collimated Consisting only of parallel rays.

The X-ray diffraction technique isn't restricted to just those substances we commonly associate with the word "crystal," such as rock candy and fine stemware. It turns out that many things in Nature can be coerced into arranging themselves into a crystal formation: proteins, DNA, other organic molecules, even entire viruses.

And crystals are easier to manipulate than their individual subunits. You can't just grab a lone molecule of, say, sodium chloride and turn it around, but you can easily pick up a salt crystal and turn around every one of its pre-aligned sodium chloride subunits all at once.

And why are X-rays preferred for examining crystals? For one thing, they can pass through substances that are opaque to visible light. But more importantly, X-ray wavelengths are just about as short as the distance between neighboring atoms in a typical crystal. Visible and ultraviolet light have much longer wavelengths. Using them to probe the fine structure of a crystal lattice would be about as clumsy as touch-typing with your fists.

Crystal Blue Precipitation

Swarovski, diamonds, Billy, Gayle — there are plenty of delightful crystals out there, but the least expensive one to surround yourself with is the one made of good old frozen H_2O. Who among us hasn't marveled at the exquisite delicacy of snowflakes? Their wide-ranging shapes are due to the very precise way in which the growth of ice crystals changes depending on the ambient temperature and humidity. But that dependence is remarkably poorly understood.

Professor Ken Libbrecht (Caltech '80) is an expert on the physics of snow: how it forms and why it takes the shapes it does. He studies it not because he wants to make a killing in winter sports or shave ice, but because of what snow formation reveals about how matter behaves at the molecular level. Here, as elsewhere in Nature, great complexity arises from the repeated application of a few simple rules.

A snowflake (or more accurately, a snow crystal) is an example of a self-assembling structure. Its six-sided symmetry is a macroscopic manifestation of the hexagonal rings into which **V**-shaped water molecules stack themselves as they crystallize (see "Hexagon with the Wind"). The reason all its arms look nearly identical is that they all grew at the same moment, in the same microclimate, their spiky shapes recording the weather's erratic changes in faithful detail.[*]

(And in response to the question you're asking right now, snow crystals are such complex objects that it's really true that — other than the simplest, tiniest, dullest ones — no two are alike.)

Libbrecht has filled several books with stunning photos of individual snow crystals. Some of his images have appeared on stamps issued by the postal services of the United States, Austria, and Sweden. But don't imagine that any old snowflake you can find will be a candidate for philatelic immortality. In the meteorological chaos that is winter weather, ugly snowfreaks are the norm, perfectly formed beauties the exception.

'S no joke.

---

*   To a good approximation, each snowflake has seven planes of reflection: six vertical, one horizontal. Theoretically, if all you had was the blueprint for one side (left or right) of one face (front or back) of one of the six arms, you could run off twelve copies and twelve mirror-image copies and assemble all twenty-four into a complete snowflake.

In What Universe?
You Can't Find City Hall

One way we might home in on the apartment building's location is by noting its relationship to various well-known landmarks. From their living room, Leonard and Sheldon have an impressive view of the distinctive red dome of Pasadena City Hall, located less than two blocks west of Los Robles

Avenue and just north of Colorado Boulevard.* The part they can see is the southeast facade, so it would appear that their North Los Robles apartment can be found south of Colorado Boulevard. (The only thing odd about that is that the portion of Los Robles Avenue south of Colorado Boulevard is called, not surprisingly, South Los Robles Avenue.)

At one point, Sheldon (as the Flash) makes an imaginary sprint from the apartment to the Grand Canyon and back again.[1] Early in his travels, he crosses Colorado Boulevard traveling north on Raymond Avenue, having evidently made a wide loop around City Hall Plaza. Doubling back onto Colorado and swinging far to the west, he heads back in toward the center of the city along the historic Colorado Street Bridge before striking out across the desert, heading westward — which we know because the January sun, high in the southern sky, is to his left. (We pay attention to these things.) It's a curiously circuitous beginning for a trip to Arizona, which lies to the east.

Or perhaps it's just his way of avoiding rush-hour traffic.

---

* Pasadena City Hall is not a part of Caltech, despite Emmy-winning cinematographer Steven V. Silver's comment to that effect in the season 5 DVD bonus features.

---

1. "The Justice League Recombination" (Season 4, Episode 11)

NINETEEN

RO-, RO-, RO- YOUR BOT

Penny: Okay, I get it, I get it. You are an emotionless robot.
Sheldon: Well, I try.
— "The Hofstadter Insufficiency" (Season 7, Episode I)

The oddest couple at Stuart's Halloween costume party must be Raggedy Ann and Raggedy Android.[1] Amy, in costume as the classic ragdoll, is over the moon just to be seen with a real live boyfriend, while Sheldon, dressed as C-3PO from *Star Wars* with a Raggedy Andy wig and cap, longs for the day when he can quit this defective carbon-based existence and store his consciousness in an everlasting silicon matrix.

He knows there are still many unsolved problems in robotics, including how to build robots in human form — androids — and how to give automata true autonomy so that they will do what we want instead of only what we tell them. He knows that many of the problems of artificial intelligence have been solved in recent years but that a full solution is still decades away.* And even when we finally do have computers that look like humans and exhibit human understanding, will we want to do away with the many foibles and weaknesses that, some would argue, actually make us human? The result could be a race of the most inflexible, insensitive, flavorless, and utterly insufferable beings imaginable.

It's likely that a completely logical intelligence, devoid of emotion and whimsy, would be not merely dull but dreadfully unpleasant to be around. That's certainly the case with humans. Amy and Sheldon can respond to each other's conversational gambits with a simple "That doesn't interest me" — a straightforward declaration of fact — and not cause offense, but most of us would read a response like that as a frosty brush-off.[2]

Sheldon's science-fiction hero, Mr. Spock from *Star Trek,* frowns upon human irrationality and has a tendency to label self-contradictory, paradoxical, or emotion-laden information or behavior "illogical" (and — when he feels it necessary to qualify an absolute — "*most* illogical"). But illogical doesn't mean irrational. Irrational thoughts and behaviors can still follow all the rules of logic (especially the logic of the

---

* Software development frequently takes longer than expected, hence Bell Labs programmer Tom Cargill's extension to the Pareto principle (the "80–20 rule"): "The first 90 percent of the code accounts for the first 90 percent of the development time. The remaining 10 percent of the code accounts for the other 90 percent of the development time."[3]

heart), while *illogical* implies misuse of those rules (see "Nuh *Uh*!"). For Sheldon, the distinction is clear: when at one point Penny asks whether he has chosen not to contradict Amy out of respect for her feelings, he retorts, "What am I, a hippie at a love-in? No. The problem is, she laid out a series of logical arguments that I couldn't refute."[4]

It's easy to view Spock's and Sheldon's emotional disengagement as intentionally supercilious. Humans are conditioned to interpret a lack of empathy as judgmental and dismissive. Think of how irritated you can get at your computer, or at a call center's hold music, for remaining so utterly unappreciative of your frustrations and deadlines. Those machines, of course, neither know nor care what's happening in your world. But whether or not you believe that computers can think, computer pioneer Alan Turing predicted in 1950 that by the end of the century we'd be speaking of them as though they could.

Turing didn't propose to establish whether a computer truly does or doesn't think. The essence of thinking remains a philosophical matter, on a par with defining consciousness. You can't read anyone's mind but your own (and you don't even know how you do that), so the only way you can draw conclusions about the thought processes of others is by observing their behavior. But personal bias gets in the way. In the 1960s, a program called ELIZA fooled a number of people into believing it was an actual person. It could conduct "conversations" by accepting typed messages and simply parroting back fragments of them in an encouraging tone, more or less as a Rogerian psychotherapist might: "I am sorry to hear you are depressed," "Tell me more about your family," "Why do you say that?" and so forth. The computer had nothing resembling awareness and was surely not thinking, but users interacted with it as though it were.

In making his prediction, Turing specifically had in mind the conversational activity now called the Turing test, but in a looser sense, it took less than ten years for his prediction to come true. Computer users commonly anthropomorphize both software and hardware. Your TV remote "gets confused." Your printer "gives up." Your laptop "hates you."

Turing test A metric for artificial intelligence in which a person conducts conversations by keyboard: one with another person and one with a computer. There are various interpretations of the test, but one is that if 70% of users can't tell which of the two was the human after five minutes of conversation, then it's fair to speak of the computer as "thinking."

On the other hand, a computer programmed to exhibit all the unpredictability and inconstancy of a human could be a maddening conversation partner. And a computer stuck on the horns of a logical paradox could be an outright menace.

The Art of Not Not-Knowing

A conversation between Leonard and Alex, Sheldon's assistant, reveals a case of either logical irrationality or rational illogic. On the subject of how often Leonard gets hit on by women, Alex playfully murmurs, "I bet it happens more than you realize." Leonard rejects this notion so dismissively — "Trust me, it doesn't" — that she backs down.[5]

Yet how can he make such a bold claim? If he's taking *realize* to mean "recognize *at the moment it happens*," then his statement amounts to "I can always tell when I'm being hit on," which, though perhaps logically defensible, is irrationally hubristic (and demonstrably wrong, as this very conversation shows). But if *realize* means "become aware of *at some point*," as Alex seems to intend, then Leonard has latched onto a bland tautology (along the lines of "I don't think I get hit on more than I think I do") and is jumping from it to an illogical conclusion ("and therefore I absolutely never get hit on more than I think I do"). Either way, he's sidestepping reason or logic or both.

Or perhaps he's just being coy.

Some paradoxes may be unavoidable. Philosopher Raymond Smullyan has pointed out an intriguing one that lies within each of us and is based on the fact that your mind is a treasure trove of ideas and opinions. Some of those are things you aren't 100% certain of, but there are plenty of others you believe to be true and treat as more or less factual: the Earth is round; humans descended from apes; convicted murderer Lizzie Borden "took an axe / And gave her mother forty whacks"; and so on. Yet each time you learn that one of your "facts" is mistaken — the Earth is actually very slightly pear-shaped (see "'Round and 'Round"); apes, like humans, descended from a common ancestor; Ms. Borden was acquitted (and it was her *step*mother, and she was struck fewer than twenty times) — you must (perhaps grudgingly) discard it from the collection of things you believe.

How then would you describe this internal collection of Things-I-Believe-Are-True? On the one hand, you believe that every item in your collection is true. On the other hand, you know that some of them are undoubtedly not true; you just don't know which ones. The collection contains no untruths, and yet it contains some untruths. So . . . does it or doesn't it?[6] Or as Sheldon snorts, "Don't you think if I were wrong I'd know it?"[7]

Smullyan's resolution of the paradox is simply to accept it as a necessary facet of human nature: we act as though all our beliefs are correct, fully aware that some of them aren't. Unless a person is terminally conceited or has some other serious psychological disturbance, he simply lives with this logical conflict, accepting corrections as they come along and taking refuge when necessary in the words of Bogart from *Casablanca*: "I was misinformed."

A similarly rational approach to the paradox of unknowability motivates Hofstadter's Law of project planning, promulgated not by Leonard but by cognitive scientist Douglas Hofstadter (who took his surname from the same person Leonard did):[8]

Hofstadter's Law It always takes longer than you expect, even when you take into account Hofstadter's Law.

Logical conflicts of this sort can result in a serious problem: cognitive dissonance. Unpleasant for humans, it's potentially disastrous for computers and robots. It turned HAL homicidal in *2001: A Space Odyssey.* It may be what makes Sheldon twitch.

cognitive dissonance That feeling of unease caused by the realization that you're holding two mutually exclusive viewpoints at the same time. The judgmental words "should" and "must" are often involved (see "Past Performance Is No Guarantee"), as are guilt and self-disgust.

For example, surely a sly fox like you can figure out a way to get those grapes; yet you've been unsuccessful thus far, so obviously you can't. The cognitive dissonance in this instance comes from simultaneously thinking, "Yes, I can solve this" and "No, I can't solve this." Another example: you hate yourself for hurling your laptop against the wall, but it felt so good at the time.

Humans have evolved various ways of dealing with cognitive dissonance: rationalization, changing the subject, pretending not to understand the question. F. Scott Fitzgerald took it as a challenge: "[T]he test of a first-rate intelligence is the ability to hold two opposed ideas in mind at the same time, and still retain the ability to function."[9]

But ask a computer in any 1960s science-fiction film to maintain two conflicting logical positions simultaneously, and it'll start smoking or spouting gibberish. This is humorously called "executing the Halt and Catch Fire instruction." In less extreme cases, the device utters, with just a hint of contempt, "That . . . does not . . . compute," a phrase first heard on the 1964 sitcom *My Living Doll.* It was spoken by an android who was also prone to concluding long utterances with a weary "This . . . is a recording."

No one, not even a robot-wannabe like Sheldon, can ignore the paradoxes of the human condition, nor should he want to. The gray area between the known and the unknown isn't merely a part of life; it lies at the heart of learning. Justice Oliver Wendell Holmes Jr. wrote, "When I say that a thing is true, I mean [only] that I cannot help believing it."[10] The very goal of scientific investigation, one might say, is to chip away from the monolith of belief every granule revealed to be false, and the reason science papers and lab reports are so fastidiously detailed is to arm the reader with as precise a chisel as possible.

"I Would Ask If Your Robot Is Prepared to Meet Its Maker ... "[10]

Caltech professor Melany Hunt has served the Institute in a variety of roles, including appointments as vice provost, vice chair of the faculty, and executive officer of mechanical engineering. For the Solar Decathlon, a biennial nationwide engineering competition, her students have designed and constructed homes that are entirely solar powered.

She's also been one of the instructors for Caltech's popular "Engineering Design Laboratory" course, in which teams of students design and build working robots, with the final exam consisting of a machine-against-machine tournament. The students earn points not only for victory in the arena but also for keeping accurate records throughout the development process and staying under budget.

Less fierce than a to-the-death robot free-for-all, the competition is structured as a race to gather and redistribute common objects, both with and without operator guidance. A robot might find itself scrambling over rocky terrain on a quest for recyclable trash, or towing a load of ping-pong balls across a campus reflecting pool. The main objective is to impose order, thus decreasing the overall entropy (disorder) of the system. However, the robots (being robots) frequently sustain a bit of inadvertent damage from one another, and at times their own personal entropy can rise significantly ... or catastrophically.

ASK AN ICON: Julie Newmar

Julie Newmar, who played *My Living Doll*'s Rhoda, a robot in the form of a large woman of spectacular physique,* went on to star as the original Catwoman in the *Batman* TV series, in a curve-following costume now at the Smithsonian Institution. Of all the actresses who have played the felonious feline fatale on screens large and small, Newmar remains a consistent crowd favorite. Sheldon makes this exact point (repeatedly) to Howard in the season 2 opener.[12]

Q: How does it feel to be Sheldon Cooper's favorite Catwoman?

Julie Newmar: Naturally, I'm tremendously flattered! Then again, everybody knows Sheldon's a robot fanatic, so perhaps he only has a soft spot for me in his heart because he's remembering me as Rhoda.

(Or perhaps he's finally perfected time travel. Following her January 1962 social visit to the Caltech campus, a photograph of Miss Newmar with four somewhat ill-at-ease male admirers appeared in the student newspaper, *The California Tech*. The similarities between that photo[13] and the cover of the *Big Bang Theory* season 3 DVD[14] are simply uncanny.)

---

*   In a tongue-in-cheek article analyzing Rhoda's

design, science fact-and-fiction writer Isaac Asimov protested: "But you *can't* build a robot in the form of a large woman of spectacular physique. That's poor robotic engineering"![15]

1. "The Holographic Excitation" (Season 6, Episode 5)
2. "The Roommate Transmogrification" (Season 4, Episode 24)
3. Jon Bentley, "Programming pearls," *Communications of the ACM* 28, no. 9 (1985): 896–901.
4. "The Spoiler Alert Segmentation" (Season 6, Episode 14)
5. "The 43 Peculiarity" (Season 6, Episode 8)
6. Raymond Smullyan, *What Is the Name of This Book?* (New Jersey: Prentice Hall, 1978).
7. "The Jiminy Conjecture" (Season 3, Episode 2)
8. Douglas Hofstadter, *Gödel, Escher, Bach: An Eternal Golden Braid* (New York: Basic Books, 1979).
9. F. Scott Fitzgerald, "The crack-up," *Esquire* (Feb. 1936).
10. Oliver Wendell Holmes Jr., "Ideals and doubts," *Illinois Law Review* 10 (May 1915): 1–3.
11. "The Killer Robot Instability" (Season 2, Episode 12)
12. "The Bad Fish Paradigm" (Season 2, Episode 1)
13. http://caltechcampuspubs.library.caltech.edu/607/1/1962_01_25_63_14.pdf#page=3
   shortcut: http://DaveZobel.com/-bbjc
14. http://ecx.images-amazon.com/images/I/91GriRcVUIL._SL1500_.jpg
   shortcut: http://DaveZobel.com/-bbs3
15. Isaac Asimov, "Why I wouldn't have done it this way," *TV Guide* 13, no.3 (16 Jan. 1965), reprinted as "How Not to Build a Robot," in *Is Anyone There?* (New York: Ace, 1967), 303–307.

TWENTY
NIGHT FISHING

Another seemingly off-the-wall science reference from *The Big Bang Theory*? Actually, the concept of luminous fish is based on real-world research. In 1999, Dr. Gong Zhiyuan of the National University of Singapore (yes, Singapore — at press time still several thousand miles from Japan) used gene splicing techniques to insert a gene for fluorescence into zebrafish embryos. (Since a cell's genes act as the blueprints for all the proteins it produces, genetic manipulation is one way to trick an organism into making molecules it otherwise wouldn't.) The fish that developed were brightly colored from snout to tail, especially under those bluish lamps commonly found hovering above household aquariums.

Gong had extracted the gene from the DNA of a jellyfish that uses it to generate the green fluorescent protein GFP (see "Hexagon with the Wind"). (Why — you may well ask — would a jellyfish need to fluoresce green? So far, only the jellyfish knows.)

zebrafish A tropical striped fish widely found in research labs and home aquariums (see "I Am the Very Model of a Modern Model Organism").

The scientists who had originally isolated the gene received the 2008 Nobel Prize in Chemistry. Gong, for his part, simply hoped to use the protein as one step toward creating a fish that could indicate when its environment was becoming excessively polluted. The idea was that pollution in the water would trigger a reaction in the fish's body that would cause it to produce more GFP, sending up the piscine equivalent of a distress flare. But he quickly realized that — pollution or no pollution — a little mint green fish makes for a darn cute pet.

Shortly afterward, his lab successfully implanted genes coding for other vividly fluorescing proteins into zebrafish embryos: a red protein from coral, an orange one derived from mutating the jellyfish gene, and even blue and purple varieties.

The resulting collection of fishy Froot Loops is now marketed under the brand name GloFish. The world's only publicly available genetically modified pet, they make a colorful (if somewhat alarming) display, reminiscent of a gaggle of aquatic bridesmaids who all somehow failed to receive the memo.

There's an important distinction between Sheldon's proposed luminous fish and Gong's real-life fluorescent ones. Fluorescent objects, like optic yellow tennis balls and high-visibility protective clothing, require an external light source (see "Better Lighting through Chemistry"). Luminous objects, like stars, glow sticks, and candles, produce their own light through chemical or nuclear processes.

It's easy to see why living things hanging out in the depths of the ocean or the dark of night might want to carry a built-in flashlight (called bioluminescence). Examples include fireflies, anglerfish, and the luminous plankton that accompany Leonardo DiCaprio on his nocturnal swim in *The Beach* (incorrectly referred to as "phosphorescent algae" in *Apollo 13*). But why an organism would need to *fluoresce* isn't always clear. In some creatures, fluorescence may be no more than an artistic side effect of the underlying biological structure. Why would humans need their fingernails and teeth to glow brightly under black (ultraviolet) light? Other than making it easier for our hominid ancestors to locate one another in prehistoric discotheques, it's doubtful this feature ever conferred any evolutionary advantage.

> **bioluminescence** The ability of certain organisms to produce light.

The crystal jellyfish, which also goes by the tuneful name of *Aequorea victoria,* is both bioluminescent and fluorescent. A bluish light produced by a biochemical reaction illuminates the animal's body, causing its GFP to glow green. This is the creature from which GFP was originally extracted.

> **California Penal System, Here I Come**
> Close to home, Sheldon is going to have a problem selling his genetically modified fish. California law severely restricts the sale and ownership of transgenic organisms (those with genes from other species); in fact, it's the only state where you can't buy, sell, or own a GloFish.

At the time the episode was broadcast, only a handful of sea creatures were known to fluoresce, but in recent years the phenomenon has been observed in nearly 200 species. What do they know that we don't know? At moderate ocean depths the faint sunlight filtering down casts a monochromatic pall of blues and blacks. The ability to glow red, green, or yellow may allow fish to identify one another or to hide in plain sight against the background of various corals and algae that, for their own reasons, are also busy fluorescing.

But they depend on that external light source. Unless a light is shining on them, a GloFish and a regular (wild) zebrafish look pretty much identical: black, with black stripes, on a black background. That's evidently not the case with Sheldon's fish nightlight; though he doesn't describe its workings in detail, what we see in the last shot of the episode is a glowing fish in a darkened bedroom. Evidently the creature is emitting light without outside help, either via phosphorescence (in which absorbed light energy is gradually released) or else via bioluminescence. (Or, just possibly, by being an ordinary goldfish under a very bright spotlight.) That's one bright little glowfish, and whatever process is lighting up its life, it's not just fluorescing.

Next question: is it *too* bright? Based on what we see in that last shot, Sheldon seems to like keeping his nightlight pretty close. He knows that light, like gravity, follows an inverse-square law; its intensity falls off with the square of distance. At twice the distance from the light source, the light is only a quarter as intense (see "The Gravity Situation").

Moms know this too, and that's why they keep telling you that if you insist on sitting so close to that TV, you'll go blind. (It's also why Sheldon gleefully tells Professor Proton, "I can get as close to you as I want without my mom saying it's going to ruin my eyes."[1]) But by that logic, any light source would blind you if you sat close enough. Not bright enough? Just halve the distance; the intensity goes up by a factor of four. Still not bright enough? Halve *that* distance. And so on.

How close to his TV would Sheldon have to sit to blind himself? How close to his fish nightlight? If he moved the bowl a few inches closer to his bed, would the glare keep him up all night? (To say nothing of what the glare must be doing to the poor fish itself. Cursed with what we can only assume are light-emitting retinas, it presumably sees nothing but a bright golden haze no matter where it looks.)

Consider this: you're in a spacecraft, so close to the dazzling full Moon (which is bright, but not blinding) that when you look at it, it fills your field of view. Now go half the remaining distance to the Moon. Thanks to the inverse-square law, the Moon is now flooding you with four times as much light as before. But it's now twice as wide (and tall) as your field of view, which means you can only see a quarter of it at a time. Even though four times as much light is reaching you, only a quarter of that light gets into your eye. You're seeing the same amount of light from one quarter of the Moon as you saw before from the whole Moon.

Now travel half the remaining distance. You're bathed in sixteen times as much light as before, but the Moon is four times the width and four times the height of your field of view, so only one-sixteenth of that light enters your eye. Again, your eye takes in no more light than before.

Peering down a drinking straw at a light bulb gives the same effect. At any distance, the increases in light and size exactly cancel each other out, and the overall brightness doesn't change. Knowing this, you can work out how brightly the ground would shine up on you if you were an astronaut on the Moon: about as brightly as our sky would shine down on us if it were crammed full of full Moons. Plenty of light, but not blinding. And that's why mashing your face into your TV (or your fishbowl) won't hurt your eyes, only your nose.

EUREKA! @ CALTECH.EDU

Biofilm Festival

Caltech's Dianne Newman is a molecular geomicrobiologist (now *there's* a mouthful!): someone who studies the relationships between microorganisms and the Earth, including how bacteria form minerals, how they dissolve them, and how the developing Earth influenced their evolution. These aren't just trivial questions. Decoding the content and history of these little critters' ingeniously complex molecular cookbooks is an important first step in understanding such questions as how photosynthesis evolved and the ongoing effects of microbes on the chemicals in the Earth's crust.

Bacteria build different proteins for themselves when their environment changes and during specific growth phases, and gene splicing is one of the techniques Newman uses to see what they're up to at any given moment. She links genes for fluorescent proteins to other genes in the microbe's DNA. That way, even if the change in the bug's activity is subtle, the change in its appearance is not.

In 2012, one of Newman's grad students, Nate Glasser, mentored a group of undergraduates who were developing a truly novel application of gene splicing. As part of the annual iGEM (International Genetically Engineered Machine) competition, students from Caltech and the California Institute of the Arts (CalArts) collaborated on a "bacterial animation" process. The "movie screen" was a Petri dish smeared with *E. coli* bacteria. The bacteria had been genetically modified by the students to sense green light and to produce a red fluorescent protein called mCherry in response. A separate genetic modification made the protein self-degrading, so that its red color faded out quickly.

In theory at least, a green moving image could be projected onto the Petri dish, and the bacterial biofilm would replicate it in vivid red in approximately real time. Though not the most convenient way to watch your favorite videos, this living screen would have an image density of up to 100 megapixels per square inch — far greater than existing films and digital devices.

And why would we need this? Because — like GloFish — why not?

OUT TO LANDS BEYOND

"One reviewer [of the paper on meringues] said the science was fine but the subject was fluffy."

Harold McGee (Caltech '73) is a world-renowned expert on the science of food and cooking. An English major at Caltech (which offers degrees in a huge range of disciplines, including the humanities), McGee went on to earn a Ph.D. in literature from Yale. He had every intention of spending his life analyzing the writings of John Keats and other literary figures, until the day he stumbled across Julia Child's dictum that meringues and soufflés are best made in a copper bowl.

Where most readers would simply make a mental note and move on, McGee went scurrying for a carton of eggs and a spectrophotometer. His analysis showed that when copper ions scraped up from the bowl's surface bind with an egg protein called conalbumin, the foam they produce is particularly stable. McGee published his findings in a peer-reviewed paper in *Nature* but revealed his literary roots by citing such authorities as James Boswell and Plato.

He now experiments, writes, and lectures full-time on food science and is the author of "The Curious Cook," a monthly *New York Times* column on the physics and chemistry of the kitchen.

---

1. "The Proton Resurgence" (Season 6, Episode 22)

TWENTY-ONE

'ROUND AND 'ROUND

Penny [watching Leonard spin an olive in an inverted drinking glass]: Wow, centrifugal force!
Leonard: Actually, it's centripetal force, which is an inward force generated by the glass acting on the olive.
— "The Fuzzy Boots Corollary" (Season I, Episode 3)

Ooh, spicy talk for a first date! Before things get much more intimate, let's clear up this centrifugal/centripetal thing once and for all, since it's been causing confusion for centuries and is apparently going to be key to Leonard and Penny's future relationship. (Glasses and olives, huh? Leonard, you dawg!*)

A rotating object "feels" as though it's being pushed

---

* The same dawg who makes the stunningly mendacious claim that his lab is "training dogs to operate the centrifuge . . . for when they need dogs to . . . operate the centrifuge . . . for blind scientists."[1]

away from the center it's spinning around. You notice this when you're watching a kid trying not to fall off one of those playground spinners. You notice it a little later when you take a curve a bit too fast and feel your shoulder pushing against the seatback as you're driving that kid to the hospital. That "feel" is often casually referred to as centrifugal force, from Latin roots meaning "center-fleeing" (*fug* as in "fugitive").

A centrifuge is a device designed to spin rapidly. To see one in action, just drop in on your local neighborhood science laboratory or astronaut training center. Failing that, you could just find a laundromat: during its spin cycle, a washing machine is a form of centrifuge. The rapid spinning causes water to leak out through small holes in the drum, while the clothes (minus one sock) remain behind.

Centrifuges are useful tools for separating things: blood cells from plasma, loose change from pockets, breakfast from astronaut candidates (as Howard discovers). Everything loose feels a push toward the outside. The faster the spin, or the more massive the object is, the stronger the push it feels. Because of this, centrifuges have been used to create a sort of artificial gravity, as much as two million times stronger than Earth's gravity.

But when Penny mentions centrifugal force, Leonard corrects her and says it's centripetal force (*pet* meaning "seeking," as in petition, but not *petal* meaning "flower part"). That sounds like it's something that forces objects *toward* the center. When would you ever see this in real life?

You see it in that veering car, when it goes around a corner and pulls your body with it. You see it in the firm grip of two figure skaters performing a death spiral. You see it in those "gravity well" demonstrations in the lobbies of science museums, where coins orbit the inside of a large bowl, whirling around and around before finally dropping through a hole in the center and landing in a trustee's pocket. All of those curving motions happen because something moving in a straight line gets pulled into a curving path instead. In fact, there's no such thing as centrifugal force; it's *all* centripetal force.

Imagine you've tied a long piece of elastic cord to a shoe and are swinging it around and around your head. The shoe travels in a circle, but if you let go of the cord, the shoe will fly off in a straight line. Thanks to Sir Isaac Newton, we know that anything that's already moving will continue moving, at the same speed and in the same direction, until a force acts upon it. Though the whirling shoe's speed never changes, its direction is changing all the time, because no matter where the shoe is in its orbit, the pull you're giving it (via the cord) is always directed toward the center of the circle, where your hand is. Your sideways pull causes the moving shoe to veer slightly to the left (or right) and trace out a curved path. Repeat this over and over, and the constantly veering shoe will have turned through a full circle.

The moment you let go of the cord, you're no longer applying that centripetal force, and the shoe continues moving along whatever direction it last happened to be traveling in.

If you spin the cord faster, your hand must exert a stronger pull to keep the shoe from flying off. As a result, the elastic cord will stretch until it has tightened enough to balance the stronger pull. Now the shoe is orbiting in a larger circle. You can continue increasing the speed, and the shoe will move in larger and larger circles until eventually the pull your hand must exert on it via the cord is so great that either the cord breaks or your willpower does.

The same thing happens if you let out more elastic cord without changing the shoe's speed of revolution. Since the circle is bigger, the shoe has to cover more ground per revolution in the same amount of time, so it moves faster. This gives it a stronger tendency to fly

Orbits

Anything in orbit must be in constant motion. As an astronaut orbits a planet, his straight-line momentum is constantly working to increase the distance between them, while at the same time gravity is constantly working to decrease it. In any given amount of time those two changes in distance more or less balance each other out. The result is that he never flies away from the planet, never crashes into it, but just continually loops around it — and that's what an orbit is (see "KISS and Tell").

If a stray chunk of rock or metal or ice hurtling through deep space passes close enough to the Sun, gravity can change its direction of travel. If the chunk is moving slowly, it may be pulled into such a tight curve that it falls into the Sun; if it's moving fast, it may veer slightly and just continue away in a new direction; but if its motion is just right, then it will go into orbit, constantly "falling" around and around the Sun, never escaping it, never crashing into it. This happens all the time, with the result that a lot of chunks of stuff are now orbiting the Sun (see "Above the Belt").

Orbits represent a balance between a moving body's straight-line momentum and a centripetal force. If you could somehow "turn off" the Sun's gravity, every orbiting chunk would head off into space with whatever speed and direction it happened to have. Just as the only thing that keeps the shoe from flying away is your pull on the cord, the only thing that keeps those chunks in orbit is gravity. Both are centripetal forces.

off, and again, your hand must exert a stronger pull to keep it from doing that, which causes the elastic to stretch and tighten a little more.

Instead of a shoe on an elastic cord, consider a ball spinning on its own axis. Each bit of the ball feels a centripetal force — it must, because any piece not feeling such a force would fly off in a straight line. The centripetal force points inward: not toward the center of the ball, but toward the nearest point on the axis of rotation. (This direction is the one that, on a planet, lies parallel to the lines of latitude. Anything at the equivalent of, say, thirty degrees north latitude is pulled directly toward the point that lies halfway between it and whatever's at thirty degrees north latitude on the opposite side.) As with our shoe on a string, the centripetal force gets stronger with increasing distance from the axis. Near the top and bottom of the ball, and at points deep within it, the centripetal pull is very weak; near the ball's equator it's much stronger.

The Earth is a little pudgy around the Equator (aren't we all). This happens for the same reason that when we spun the shoe in a wider circle, our elastic cord stretched. Although the planet rotates at a uniform rate, how fast any point on (or in) it moves sideways through space depends on how far away from the axis that point is. A piece of material close to the North Pole, for instance, has a low sideways momentum, which the combination of Earth's gravity and stick-together-ness are more than enough to counteract. The result is that the North and South Poles are pulled toward the Earth's center.

Near the Equator, where the surface is so much farther out from the axis, it's moving much faster. At these higher speeds, the planet's gravity and stick-together-ness aren't enough to overcome the straight-line momentum of its components, and the surface moves outward a little, just like the shoe on the elastic cord. This motion stretches the Earth — tightening the rubber band, as it were — until the additional centripetal force supplied by the stretching just balances the material's tendency to fly off. The overall effect is barely noticeable; the straight-line distance across the Equator (roughly 8,000 miles) is only about twenty miles (or a quarter of one percent) longer than the distance from pole to pole.**

The equatorial bulges of those giant starry pancakes called galaxies are much more obvious. A spiral galaxy starts life as a spinning ball of gas. It has a much greater mass than the Earth and a much more impressive gravitational pull, but it's much less dense, so it's more stretchable and collapsible. (Its components are too widely separated to stick to one another the way the components of planets do.) Everything in the galaxy is being pulled directly toward its center by mutual gravitational attraction. This is not some mysterious property of galaxies but is a result of the combined gravitational attraction of

---

** In fact, the Earth is more nearly spherical than the international manufacturing specification for billiard balls allows, giving rise to Archimedes's oft-misquoted claim that if only someone would give him a long enough stick, he could pull off one heck of a bank shot.

all the individual bits of matter the galaxy is made up of. No matter where you are in it, if you were to face toward the center, there would be more material in front of you than behind you: more pulling you forward than backward. That's why gravity always points toward the center of a galaxy or the Earth or any reasonably uniform and symmetrical object.

The momentum of any object within the spinning galaxy is a straight line pointing sideways, but gravity's pull toward the center of the galaxy bends the object's path in two directions simultaneously: toward the galaxy's axis, and toward the circular region bounded by the galaxy's equator. The former pull is the centripetal force that keeps the object in orbit, but there's nothing to counteract the latter pull, with the result that anything above or below the galactic equator migrates toward it unopposed, and over time the galaxy flattens out into a pancake shape.

On a smaller scale, the same thing happens when a solar system is born. A spinning blob flattens out into a star surrounded by a disk of loose material that eventually clumps up into a bunch of planets, all generally revolving in the same direction and in more or less the same plane. Planets that don't toe the line in this regard are probably later additions, as is the case with Pluto (see "The Naming of Things").

The reason a disk of dough being spun by a pizza chef flattens out is related, but the forces are different. The gravitational attraction of even the largest pizza crust is so tiny it's irrelevant, and the centripetal force in this case is supplied entirely by the stretchiness of the dough. Just as with the shoe on the elastic cord, the

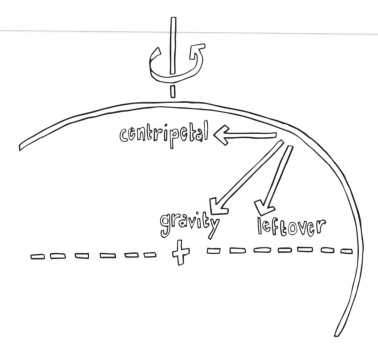

Cross-section of a spinning galaxy collapsing under its own gravity. Part of the gravitational pull provides the centripetal force that keeps objects orbiting, and the remainder pulls them into a pancake shape (the dashed line) rather than directly toward the center point (marked with a +).

outermost (fastest-moving) edges stretch outward until the dough becomes firm enough to apply a centripetal pull that exactly balances the edges' straight-line momentum.

In a moving car, your body will continue traveling in a straight line until a force acts upon it. Turning the car to the right doesn't immediately do anything to your body — it simply moves the left side of the seatback into your left side. The resulting pressure causes your body to move to the right. Since your straight-line path has veered slightly toward the center of a circle (your turning circle), this was a centripetal force. It may feel as though your body is applying a force of its own, a centrifugal force that pushes outward against the seatback, but that's an example of a fictitious force: one that is seen only when the observer is accelerating (experiencing changes in direction and/or speed). To see that a centrifugal force is a fictitious force, imagine standing on a block of ice on top of the car (not recommended). As the car turns underneath you, it applies a centripetal force to its occupants, but it has no way of applying one to you, so rather than turning with it, you'll continue moving in a straight line. For a time, anyway.

You Only Think You're Falling
Gravity, too, can be viewed as a fictitious force. An object with mass creates a warp in space-time that causes other objects to move inward not because they're being "pulled" but because a straight-line path passing through warped space-time bends into a curve.

Unlike real forces, fictitious forces can be made to disappear by switching to an inertial (non-accelerating) reference frame. Leonard starts to explain this to Penny just before he passes out (oops — spoiler alert). For instance, the occupants of the cornering car may feel something invisible pushing them to the left, away from the turn, but up on your block of ice you recognize that the car is pushing them to the right, off their straight-line paths.

Where did the notion of centrifugal force come from? Probably from the fact that, as with most of our life experiences, we view our own reference frame as the most convenient description of events (see "Four and a Half Lives"). It's typically easier to see things from your own point of view than from someone else's, no matter whose reference frame is simpler or otherwise "better." We also tend to focus on the positions of objects rather than their velocities, although forces are associated with changes in velocity.

Moreover, the word *centrifugal* is well established in the argot. In another episode, Leonard contemplates making the ultimate sacrifice (for him, anyway) just so the physics department can get its hands on a device called a cryogenic centrifugal pump.[2] This is a machine that spins like an electric fan, drawing liquid in, whipping it around in a circle, and shooting it out through a tube. Cryogenic just means it's for pumping very cold liquids, but centrifugal is a holdover from earlier centuries, when engineers used to sling words like that around because they didn't have pedants in glasses (with olives in their glasses) to correct them.

So now that we know all that, let's get back to the vital question: How *does* centripetal force keep the olive in the glass? The answer comes to us no thanks to Leonard, whose bare-bones explanation no doubt is giving Penny the impression that centripetal force is something that makes olives float in the air.

The actual process, though somewhat tedious to describe, is easy to perform once you've acquired the knack. In brief, the centripetal force supplied through careful and constant adjustment of the glass only keeps the olive moving around the glass in a circle, rather than flying off in a straight line. The thing that allows the olive to rise above the table is a second, vertical force — one that counteracts gravity — which Leonard continually supplies by re-angling the glass ever so slightly, this way and that, so that the spinning olive tends to roll very slightly upward from time to time. Without a fairly steep-sided glass, an olive that rolls well, a little bit of friction, and those constant angular adjustments, his plan to get a lapful of Penny would have ended with nothing but a lapful of olive.

EUREKA! @ CALTECH.EDU
Win One for the Zipper

When a DNA molecule replicates and becomes two, what exactly happens to the original molecule? Does its famous double helix remain intact, acting as a sort of template for the new molecule, as though photocopied? Is it chopped up into fragments that are reassembled with new fragments to make two new molecules striped like spumoni? Does it unzip down the middle, with each half getting a new half assembled onto it?

Let's say n (for normal) represents the mass of one DNA molecule in some particular bacterium. If you raise those bacteria on a diet containing heavy nitrogen (nitrogen with an extra neutron in each atom), all the DNA molecules they create for themselves will contain heavy nitrogen, so each molecule's mass will be just a bit more than n. We'll use h (for heavy) to represent the mass of those slightly heavier molecules.

If you then switch the bacteria to a normal-nitrogen diet, any new DNA molecules they assemble from then on will use those normal nitrogen atoms and will be less massive than h. But how much less massive? That depends on which model of replication is right. In our "photocopier" model (called the conservative model), every DNA molecule in the bacterium either is one of the original, heavy ones, with mass h, or is made of all-new, non-heavy nitrogen and has mass n. In our "spumoni" model (called the dispersive model), every molecule consists of a patchwork of old and new bits and has a mass somewhere between n and h. In our "zipper" model (called the semiconservative model), every molecule either contains only all-new nitrogen (and has mass n) or has inherited one of the backbones from one of the original, heavy molecules (and has a mass halfway between n and h).

Caltech's Matthew Meselson and Franklin Stahl devised and performed this experiment in 1957. They successfully extracted the DNA molecules from the bacteria, mixed them into a salt solution, and spun it in an ultracentrifuge (which is just like a regular centrifuge, only ultra). The molecules settled out into visible layers according to their mass: the heavier the molecule, the lower down it settled.

And what to their wondering eyes should appear? Some molecules had mass n, some had a mass halfway between n and h, and none had mass h. Zipper . . . for the win!

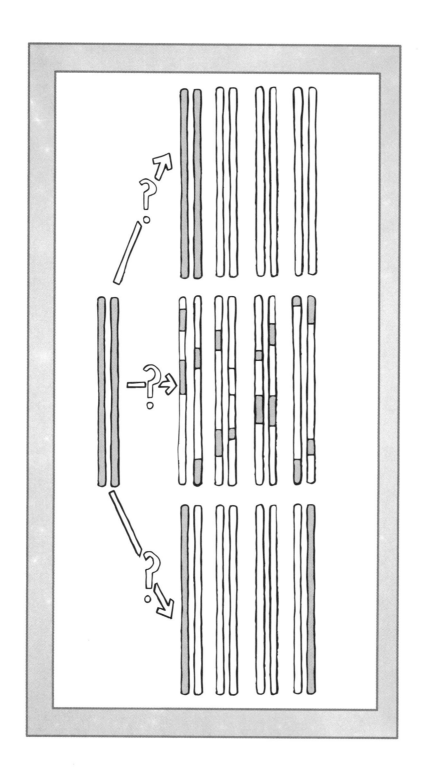

Everybody's Favorite Game Show: *Spot the Typo!*

Sorry — the subject of the lecture Sheldon refuses to co-present with Leonard is *what* exactly?[3] Inside the ballroom, one of Leonard's overhead slides says:

Moment of Inertia Measurements

Torsional Pendulum: $\tau = 2\pi(\kappa/I)^{-1/2}$

$$I_{tot} = I_{class} \times (1 - n_s(T)/n)$$

N. B., x-axis in log Kelvin

He concludes the presentation with: "[A]t temperatures approaching absolute zero, the moment of inertia changes . . ."

Moments later, Sheldon describes what first got him thinking about "the moment of inertia in gases like helium."

But the placard outside the ballroom says:

the Institute for Experimental Physics proudly welcomes
Leonard Hofstadter, Ph.D
presenting "Paradoxical Movement-of-Inertia Changes Due to Putative Super-Solids"

(Hint: Leonard obviously wasn't responsible for the placard.)

---

1. "The Pancake Batter Anomaly" (Season 1, Episode 11)
2. "The Benefactor Factor" (Season 4, Episode 15)
3. "The Cooper-Hofstadter Polarization" (Season 1, Episode 9)

TWENTY-TWO
LASER FAIR

MacArthur "genius" grant awardee David Underhill is working with Leonard on a new laser-based detector for studying dark matter. He doesn't tell Penny what kind of a laser he's using, but it probably doesn't make much of a difference (not to her, certainly), because in a lab full of cool toys, lasers "and stuff" may just be some of the coolest.

Here's a partial list of the many high-tech uses to which lasers have been put on *The Big Bang Theory*:[1]

* burning a hole in a plastic toy
* heating up soup

* burning a hole in a balloon
* knocking out incoming ballistic missiles
* studying the soft component of cosmic radiation
* front-projected holographic display with finger tracking
* burning a hole in the wall.

The beam of a laser is coherent, meaning that the emitted photons (particles of light) are vibrating in sync with one another. In the absence of coherence, some of the photons would cancel out the influence of others nearby, just the way noise-canceling headphones suppress sounds coming in from outside by adding an out-of-sync sound of their own (see "Can You Hear Me Now?").

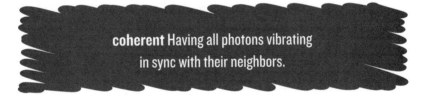

coherent Having all photons vibrating in sync with their neighbors.

Laser beams are also highly collimated, meaning that their photons travel along parallel paths, rather than diverging (see "A Thousand Points of Light"). A collimated beam produces a focused spot, suitable for delivering a large amount of energy to a small area at considerable distances. Laser applications include medicine, communications, astronomy, data processing, power management, weaponry, and dozens of other fields. There's one in every CD player and laser pointer. But how do they work?

Although all lasers use some of the radiation they produce to stimulate the production of more radiation, there are several different types, and they operate on different principles. The first ones ever invented relied on the light-emitting properties of solids, gases, and vapors, but the majority of lasers produced nowadays are based on semiconductors.

Most of the lasers Leonard works with consist of a tube with mirrored ends. The tube may contain a single type of atom or a blend. Photons from an external energy source illuminate those atoms, temporarily exciting some of their electrons to higher energy states.

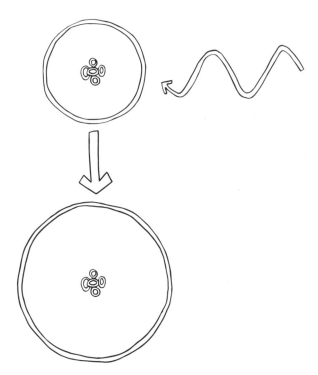

An atom absorbs a photon, bumping up the energy level of one of its electrons. No part of this picture is to be taken literally.

As each electron makes its way back to its lower energy state, it releases a photon. Since each type of atom permits its electrons to inhabit only certain energy levels and never any others, the energy of the released photon is limited to a specific set of possibilities corresponding to specific wavelengths (colors) of light (see "Better Lighting through Chemistry").

Left to its own devices, an excited electron will normally come back down to a lower state in its own good time. But if a second photon happens to strike the atom, it can knock the electron down to a lower state immediately, forcing it to emit its photon sooner than it otherwise would have. Moreover, when the new photon is released, it will be vibrating in sync with the photon that freed it, as well as traveling in the same direction.

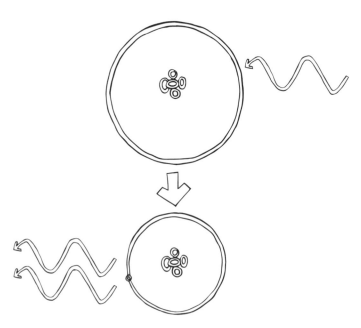

A second photon stimulates the atom to emit a photon — in sync.

This photon emission process doesn't happen just once per atom. Another incoming photon can re-excite the electron, still another can knock it back down, and so on. In a sense, the electron acts as a temporary storage and redirection facility for photons, where regardless of what the original photon was doing when it arrived, both photons will be traveling side by side and in sync when they depart. (It's like finding yourself singing, "Hap-py birth-day, dear Pro-fes-sor Katz-en-el-len-bo-gen" just when everyone else is singing, "Hap-py birth-day, dear Re-née. . ." The song falls into chaos and people look around in confusion, but finally someone plunges ahead loudly into the last line — at which point you all launch into song again, in unison.)

And what happens to those two photons after they depart arm in arm? They bounce off the mirrored end of the laser tube, still side by side and in sync, and travel back down the tube, drawing more photons into lockstep with themselves. The number of photons moving in sync and parallel to the axis of the tube quickly overwhelms all others, and thereafter no rogue photon vibrating out of sync or traveling at an odd angle can hope to remain that way for long.

It's all very well to have a tube full of coherent and collimated light, but we need to get access to it. This is done by making the mirror at the business end of the laser very slightly transparent: perhaps 99% of the light that strikes it is reflected back into the tube, but the other 1% leaks out, forming the laser beam.

And no, despite what you hear nearly every time someone fires one on TV or in the movies, lasers don't go zzzap or whirr or pshew or ffmp-ffmp-ffmp. Their power supplies can be noisy, but the laser itself does its work silently. For that matter, regardless of what the media would have you believe — and *The Big Bang Theory* is no exception — the beam of a laser, like that of any light source, is invisible except when striking something or passing through a medium that scatters it: smoke, fog, or olive oil, for instance. (The bright spot your laser pointer makes on the wall is almost your only indication that it's turned on, which is what makes it such an ideal device for messing with your cat's mind.) That's why Leonard has to spray aerosol disinfectant in order to reveal the positions of the crisscrossing laser beams in Secret Agent Laser Obstacle Chess.[2]

An entirely different principle lies behind the free-electron laser, the device that Leonard announces he was "up all night using" for an X-ray diffraction experiment (see "A Thousand Points of Light").[3] The free-electron laser is the most powerful mechanism ever devised for producing X-rays, which can't be generated by other types of lasers because they would pass right through the mirrors at the ends.[*] This coolest of cool toys was first conceived of by Dr. John M.J. Madey during his undergraduate days at Caltech. The powerful beam of electromagnetic radiation it

---

[*]   The physics department has a free-electron laser for Leonard
      to work with less than one season after North Korean
      wunderkind Dennis Kim pooh-poohs them for not having
      one.[4] Evidently, the budget committee — or the writers —
      were listening to Dennis Kim.

produces has its origins not in a hall of mirrors filled with atoms luxuriating in their own reflected glory, but in the hubbub of a torrent of electrons hurtling down a magnetic corridor at nearly the speed of light.

Normally the electrons in an atom, regardless of their energy states, never stray far from the positively charged nucleus on their own. With a little electromagnetic coaxing, however, one or two can be induced to leave the atom behind, becoming free electrons for a time (see "One Potato, Two Potato"), and during their all-too-brief moment of freedom they can be sent on an astounding joyride.

A free electron tends to travel in a straight line, but in the presence of a magnetic field, it can give up a smidgeon of its energy of motion. That bit of kinetic energy squirts out sideways, where it's seen as a burst of electromagnetic radiation (such as light, X-rays, or gamma rays).

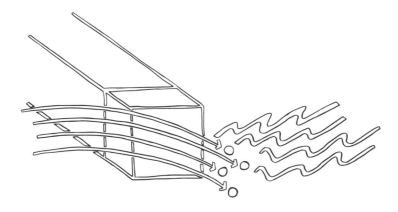

Cyclotron radiation. The magnetic field causes the electrons to give off light and change course.

Meanwhile, the recoil from the emission gives the electron a slight kick toward the opposite side, causing it to veer off from its straight-line path and follow a curve instead. It's the same thing that happens anytime Howard, floating in the International Space Station, throws something up (e.g., his lunch): his body gets a push downward.[5] Because this kind of radiation is dramatically present in the type of particle accelerator called a cyclotron, it's known as cyclotron radiation. You don't need a cyclotron, however: cyclotron radiation is released any time a charged particle moves through a magnetic field.

At higher speeds, if the particle is traveling almost as fast as the radiation it's emitting, some exciting additional effects occur. The radiation gets a sharp boost in frequency and is directed strongly forward, rather than to the side. This is synchrotron radiation, named after the device where it was first

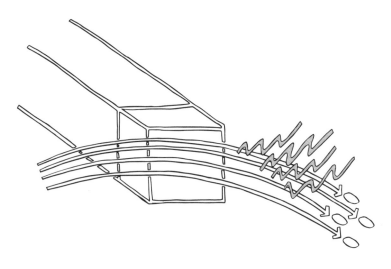

Synchrotron radiation

seen: the cyclotron's higher-energy big brother. But you don't need a synchrotron to produce it — just a free-electron laser.

Picture a channel lined with two long rows of strong magnets. North and south poles face each other across the gap and alternate down the length of the row. This assembly, called an undulator, is one of a broader class of devices called wigglers (yes, seriously).

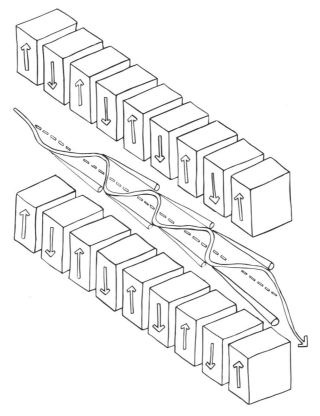

Undulator, alligator. (Wiggle a while, crocodile.)

A high-speed beam of free electrons, similar to the beam inside an old-fashioned TV tube but pre-accelerated to nearly the speed of light, is directed down the undulator channel. The push-me-pull-you force that the electrons encounter as they run that magnetic gantlet wiggles them left and right like a conga line (or, if you look at the picture sideways, up and down like a long, straight roller coaster). Wherever their path curves, they emit synchrotron radiation, and since the path is most sharply bent at the outermost extremes of the slalom, it's there that they radiate most strongly, sending forth bright beams of high-energy photons.

Those photons aren't vibrating in sync, though; they're just being emitted randomly, whenever an electron happens to go over one of the magnetic bumps. But a peculiar thing happens when photons emitted near the back of the line, racing forward as dry leaves that before the wild hurricane fly, overtake electrons up ahead. The photons push and pull the electrons into groups, giving a forward shove to some and pulling back on others. What began as a continuous stream of particles is quickly massaged into a rapid-fire series of electron bunches, spaced one photon wavelength apart. The more tightly bunched the electrons are, the more coherent will be the beam they send forward, which causes still tighter bunching of the electrons up ahead, which send forward a still more coherent beam, and so on, all the way to the front of the line.

Most lasers' energy levels are tied to the state transitions of their constituent atoms and electrons, which are hard-wired by Nature. But a free-electron laser can be tuned across a wide range of wavelengths simply by changing either the dimensions of the undulator or the speed of the electrons.

Unfortunately, they're expensive to build and operate, which is why there's presently only a handful of them in the world. Most likely, Leonard was "up all night" not because he was running an experiment that took a long time, but because he was waiting for his turn. Such is life when (according to Sheldon) you're "just a dime-store laser jockey."[6]

EUREKA! @ CALTECH.EDU
Lilliputian Lasers

A tube with mirrored ends is the most common visible-light laser configuration, but it's not the only possible one. Caltech professor of applied physics Kerry Vahala fabricates tiny O-shaped optical devices that can achieve the same effect by sending light on a whirling one-way merry-go-round ride.

Each of his toroidal microresonators ("doughnut-shaped tiny lasers" — in Danish, no doubt) consists of a silicon fiber as fine as a spider's thread looped into a circle barely big enough for a human hair to pass through. Light allowed to leak into the ring from a nearby optical fiber races around and around unimpeded, building in strength before finally escaping into another adjacent fiber. Adding the right materials to the silicon at fabrication time yields a solid-state laser that, because of its tiny size, consumes only a fraction of the power of a tabletop device.

Vahala's microresonators are so finely crafted that each time the light completes one trip around the ring, only a millionth of a percent of the power leaks away. If a bell worked that efficiently, you could strike it once and it would ring for days.

Laser luminary William Bridges is the discoverer/inventor of the ionized noble gas lasers, the first of which was the argon ion laser. A strong advocate for women in the sciences, he was responsible for reactivating Caltech's long-dormant chapter of the Society of Women Engineers (SWE) in the 1980s.

Bridges can be spotted enjoying fine cafeteria dining during season 6 of *The Big Bang Theory*.[7]

Q: Penny's momentary interest aside, women continue to be dramatically underrepresented in high-end science and engineering positions. What can you say to a young woman entering the field who looks around her and says, "Why don't I see more people like me?"

William Bridges: It wasn't always this way. In 1943 my great-aunt Mary was Rosie the Riveter, building B-25 bombers while Russian women were flying combat missions in Europe. But ten years later, when I was a Berkeley undergraduate, there were exactly zero women in my engineering classes. I couldn't understand that; engineering is so interesting!

In the early '80s, when the numbers had risen a bit, I recruited a couple of young female engineers from the Missile Systems Division of Hughes Aircraft to make a presentation to the Society of Women Engineers at Caltech. They described their missile projects with great enthusiasm — "*swoosh! boom!*" — and it was fascinating. But I realized there were many in the audience who couldn't get past the

> fact that the things these women were developing could be used to kill people. I'm not judging — it's just an observation.
>
> So we still have some hurdles to overcome. Or perhaps we're just giving girls the wrong toys.

1. "The Psychic Vortex" (Season 3, Episode 12), "The Fuzzy Boots Corollary" (Season 1, Episode 3), "The Workplace Proximity" (Season 7, Episode 5), "The Recombination Hypothesis" (Season 5, Episode 13), "The Jerusalem Duality" (Season 1, Episode 12), "The Holographic Excitation" (Season 6, Episode 5), and "The Cooper/Kripke Inversion" (Season 6, Episode 14), respectively.
2. "The Work Song Nanocluster" (Season 2, Episode 18)
3. "The Euclid Alternative" (Season 2, Episode 5)
4. "The Jerusalem Duality" (Season 1, Episode 12)
5. "The Table Polarization" (Season 7, Episode 16)
6. "The Speckerman Recurrence" (Season 5, Episode 11)
7. "The Egg Salad Equivalency" (Season 6, Episode 12)

TWENTY-THREE
ICK-ACK-SPOCK

> Sheldon, Howard, and Raj: Rock-paper-scissors-lizard-Spock!
> [All three make the Spock gesture.]
> Sheldon: Okay, one of us is going to have to stop putting up Spock.
> Howard: How do we decide that?
> [A thoughtful pause.]
> All three: Rock-paper-scissors-lizard-Spock!
> [All three make the Spock gesture.]
> — "The Lizard-Spock Expansion" (Season 2, Episode 8)

Is there anyone over the age of six who doesn't know how to play rock-paper-scissors? The rules of this schoolyard game (and its variants) are about as simple as rules can get. The players (usually two) face each other, and after chanting the game's name in unison, or "once-twice-three-shoot" or "ready-set-go" or some similar incantation (a step called "priming"), each player displays one of a set of hand gestures.

If both players have made the same gesture, then the game is a tie and generally goes into a rematch. Otherwise, one gesture defeats the other according to a specified hierarchy. The winner may acknowledge his superiority by animating his gesture in a manner suggesting the applicable rule (e.g., the "rock" fist pushes into the spread fingers of the "scissors" hand as though dulling or breaking them; the flat "paper" hand clutches the "rock" fist as though wrapping it; the "scissors" fingers close down on the flat "paper" hand as though cutting it).

The key feature of the game is its non-transitivity, meaning that everything can be defeated by something else. (Card games in which ace beats king but deuce beats ace also have this property.) Strategy, to the extent that there is any, hinges on the ability to "read" one's opponent or to influence him subconsciously . . . or, failing that, to bore him into submission.

The game isn't normally played with more than two people and three gestures, but there's no law against it, and that's what's going on in this three-handed tournament, using the five-gesture extension known as rock-paper-scissors-lizard-Spock.

Adding players and gestures doesn't reduce the chance of a tie, and Sheldon's declaration that someone is going to have to break the deadlock by not putting up the "Spock" gesture gives a clue to a possible strategy. Suppose two players plan to continue putting up Spock, while the third plans to win by putting up paper instead (which disproves Spock) or lizard (which poisons Spock). What would be a better strategy for either of the other players? Paper can be beaten by scissors (which cuts it) or lizard (which eats it); lizard can be beaten by scissors (which decapitates it) or rock (which crushes it). So if one player is likely to put up paper or lizard (to defeat the other two's Spocks), one of the other players should put up scissors, which beats both paper and lizard.

roshambo \roe-sham-BOE\ Another name for the three-gesture game. The etymology of the word is unclear; on the one hand, the similarity of the rosh- part to *roche*, the French for "rock," is interesting, but the dissimilarity of -ambo to the French for "paper, scissors" means the origin is still a matter of conjecture. Some have attempted to link the word and indeed the game to the comte de Rochambeau, commander-in-chief of the French Expeditionary Force during the American Revolution, but that's probably about as likely an origin story as the notion that Marco Polo enjoyed splashing around in kiddie pools with his eyes closed.

But this strategy fails if the third player continues to put up Spock (as Sheldon, owing to a case of almost pathological media self-identification, seems particularly likely to do), since Spock smashes scissors. So if a player thinks the other two will put up either lizard and Spock or paper and Spock, it's safest for him to put up lizard, since the first case ties with lizard and beats Spock, and the second case beats both paper and Spock. Then again, both of the other players have access to this same knowledge, and either of them may adopt the identical strategy, which would result in another tie. In the most likely case, the next round would be lizard-lizard-lizard. A player who realizes this can win simply by putting up something that defeats lizard (rock or scissors). But what if another player comes to that same conclusion? Or what if another player continues to put up Spock, which defeats both rock and scissors?

It's obvious that there's no end to the number of steps of reasoning that one can pursue in this way. Even with only two players, a reasoned argument in favor of any specific gesture could instead result in a tie (if the other player makes the identical argument) or in a defeat (if the other player takes the argument one step further).

This sort of "that's just what they'll be expecting us to do" logic occurs frequently in game theory and in the analysis of strategic games like chess. It's unofficially known as "Sicilian reasoning" in honor of its appearance in the film *The Princess Bride*, where during a duel of wits involving goblets of poisoned wine a character boasts that his superior intellect comes from being Sicilian. He then proceeds to become hopelessly entangled in a never-ending chain of inconclusive conclusions, along the lines of "ah, but of course you must have *known* I'd change my mind, so I'm now going to change my mind *again*!"

But What About Al Gore?

A few seasons after rock-paper-scissors-lizard-Spock was introduced on the show, it made another appearance. This time, Sheldon carefully attributes the game's origins to "internet pioneer Sam Kass" and hails him enthusiastically in absentia. The reason for this is simple: not only had the show's writers not invented the game, but in the intervening time period they had discovered (to their mortification) that it wasn't in the public domain. Sam Kass is a real person, and the earlier episode's uncredited use of his game's name and rules could be seen as a copyright infringement.

As it happens, neither Kass nor game co-creator Karen Bryla (who remains mystifyingly unhailed and uncredited) seemed inclined to pursue the matter. Nor, if they had, would the lawsuit likely have been resolvable by anything as simple as a game of rock-paper-scissors. But the answer created a new question: What's this "internet pioneer" label? Did Sam Kass invent the internet?

No: both the internet's structure and its content are the growing creation of a vast group of people. (That, evidently, is what *Time* magazine was getting at when it named "You" — i.e., anyone who had ever created any scrap of online content — its "Person of the Year" for 2006.) Kass and Bryla were internet "pioneers" mainly in the sense that, having created the game, that's where they published it.

This is not to take anything away from them — their clever invention has delighted millions, and as an improved method of conflict resolution, no doubt it's saved untold numbers of young nerds from poundings by schoolyard bullies. (Yes, isn't it pretty to think so?) But it's important to guard against letting our perspectives get distorted. There's enough misunderstanding about the Information Age as it is.[**] Only from the point of view of a studio's nervous legal department could the word "pioneer" be considered synonymous with "user."[****]

---

[*]   Although a U.S. federal judge did actually order this once, in connection with a different and more trivial issue (*Avista v. Wausau*, 2006).

[**] Al Gore never claimed to have invented the internet, Bill Gates is not the father of the personal computer, the NSA is not having a giggle over every text message you send, etc.

[***] Back in 1997, when a Massachusetts judge made the novel decision to distribute one of his rulings online in addition to the traditional print version, Luddites worldwide decried his action as blatantly exclusionary, many blaming the judge's "computer-crazy" son for his presumed role in promulgating such egregious techno-elitism.

With quick enough reflexes, one could in principle delay deciding on a gesture until the hands were coming down and the gesture the opponent was starting to form had become obvious. A robotic arm built at the University of Tokyo implements just this strategy and will (somewhat eerily) beat a human player every time, simply by relying on a high-speed video camera to observe the human's hand beginning to form the gesture.

Absent a technological boost of that sort, players are generally competing from symmetrical positions, in which none of them can either ascertain or influence what any of the others will do. There are strategies that attempt to break this symmetry, such as by giving or detecting subliminal cues, but in a truly symmetrical situation, the best strategy is to play completely randomly. Over a large number of games, a person who plays in this fashion is likely to lose as often as win, whereas any non-random strategy runs the risk of being discovered and exploited by the opponent.

ick-ack-ock Yet another name for the three-gesture game, more common in the UK than elsewhere. The name may be connected with the Latin *hic, haec, hoc* ("this, this, this") as spoken by a Cockney, or with tic-tac-toe, or with a cat coughing up a hairball. Or it may not.

Unfortunately, humans can never be truly random (despite what you may have observed on the highway and at election time); various influencing factors are always at play, both obvious and subtle. People have exploited this by writing algorithms that track the behavior of players and attempt to glean patterns. To play truly unpredictably, one would need to consult (and obey) a significantly random source, such as by throwing Dungeons and Dragons dice or analyzing radio static or observing the motion of subatomic particles. One can only imagine how schoolyard-bully-mollifying *that* would be.

To some, the appeal of rock-paper-scissors comes in designing variations on the theme. In principle, any number of gestures greater than two can be agreed upon, although if the number is even, then some gestures will lose to more gestures than they beat, and vice versa. This can become part of a double-bluffing strategy ("he'll never expect me to throw a gesture that loses more than it wins, so he'll feel safe throwing something it would beat, so I'll throw it and surprise him"), but playing completely randomly is no longer likely to win half the time.

One could take this to extremes, as artist Dave Lovelace has done. After a full year of pondering, head-scratching, and covering walls with sticky notes, he at last announced that he had successfully designed a 101-gesture game (no, that is not a misprint). The rock, the paper, and the scissors had been joined by a whole ark-load of nouns, including such arcana as chainsaw, cockroach, Medusa, beer, and poison. Having developed and memorized all 101 gestures and all 5,050 interactions (again, not a misprint), Lovelace encountered little difficulty in teaching it to his friends, as he no longer had any.

There's nothing magical about rocks, paper, scissors, or any of the other items in any of these games, of course — not even Spock or chainsaws. (However, by international agreement, Chuck Norris beats everything.) You could play a perfectly serviceable (if dull) version of the game by using colors or shoe designers or flavors of ice cream — provided that the gestures and the various win-lose relationships are all defined ahead of time.

Some international variants on the game involve neither rocks, paper, nor scissors. They include man-elephant-earwig (man flicks earwig, elephant stomps man, earwig tortures elephant) and tiger-village chief-village chief's mother (tiger threatens mother, chief chases tiger away, mother's nagging makes him kind of wish he hadn't). There's also a five-gesture extension of rock-paper-scissors in which guinea pig deactivates dynamite by nibbling its fuse; the other interactions are left to the reader's imagination.

Make Your Own Game

An easy way to create a rock-paper-scissors variant is to choose an odd number of interesting items and make up a gesture (fist, flat hand, etc.) for each one. Now write the names of the items in an evenly spaced ring, in such a way that each one defeats its counterclockwise neighbor but is defeated by its clockwise neighbor.

Connect each pair of items (whether or not they're neighbors) by a straight arrow. All arrows must point counterclockwise (as the diagram is rotated, any arrow whose tail is at the top must point to an item on the left side of the ring).

Finally, associate a meaningful interaction with each arrow (paper wraps rock, scissors cut paper, etc.), bearing in mind that the item at an arrow's tail beats the item at its head. This last part can take some time because the number of arrows is usually much greater than the number of items (three items need only three arrows, but five items need ten arrows, seven need twenty-one, nine need thirty-six, etc.).

For example, here's rock-paper-scissors-lizard-Spock:

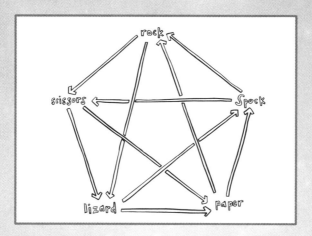

Each arrow points from winner to loser, reminiscent of the common practice of covering the loser's gesture with the winner's (and the even more common practice of pointing at losers). This is not the only way the diagram could be laid out, but it has the advantage that regardless of how you rotate it, the gesture at the top beats all gestures to its left and loses to all gestures to its right. That's not the case with the commonly seen version in which some of the arrows go clockwise and some go counterclockwise:

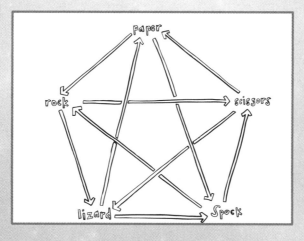

Then again, that version has the advantage that the name of the game is easy to read from left to right. But look more closely. Though the name is easy to read, it isn't laid out symmetrically around the circle. You can see this if you try to trace out the pattern from rock to paper to scissors to lizard to Spock. After going clockwise for a while, you'll suddenly find yourself skipping across the circle and then going counterclockwise.

The game's name doesn't make a symmetrical pattern in the other diagram either; lizard is out of order. In fact, by the rules of symmetry the game should really be called "rock-paper-scissors-Spock-lizard," and when Sam Kass and Karen Bryla invented it, that's exactly what they called it.

If you erase Spock, the lizard, and the seven arrows going to and from them, you're left with the diagram of rock-paper-scissors, as before. That's because the diagram of the five-gesture game contains the diagrams of all possible three-gesture subgames. Rock-paper-scissors isn't the only subgame, either; rock-Spock-lizard is another (rock crushes lizard, lizard poisons Spock, Spock vaporizes rock).

Not all the subgames are balanced, however. Rock-paper-Spock is an unbalanced game because paper never loses and rock never wins. So is paper-scissors-lizard: paper never wins and scissors never loses. So how do you easily find all the balanced subgames inside a larger game?

First, choose just some of the items. (We'll choose three. An even number of items always gives an unbalanced game, and a one-item game would end in a tie every time.)

On the diagram that contains only counterclockwise

arrows, trace out a route that connects each of your chosen items to the next one around the circle. Ignore all the other items and arrows. If you can follow the arrows all the way around this (probably lopsided) ring and return to your starting point, the game is balanced. But if any arrow in the ring points the wrong way, the game is unbalanced. In that case, swap one of your items for a different one and try again. (You can choose any item to swap out; no matter which item you delete, there's always at least one item elsewhere on the circle that will balance the game.)

For example, within rock-paper-scissors-lizard-Spock we can trace out the three arrows that make up the ring of rock-paper-scissors, and we can trace out the ring of rock-Spock-lizard, but when we try to go around the ring of rock-paper-Spock, one of the arrows points the wrong way, indicating that the game is unbalanced.*

You can name your game anything you want to, but a logical naming system would start somewhere and follow the arrows backward (clockwise) around and around the ring, bypassing a fixed number of items (zero or more) each time. By this method, rock-paper-scissors-Spock-lizard can also be called rock-Spock-paper-lizard-scissors or paper-lizard-scissors-rock-Spock or any of seven other names.

Having peered into the game to find some smaller games, you might like to try to build it up to make some bigger ones. An easy way to do this is to add items two at a time, as follows:

Invent two new items, plus a gesture for each. (We'll call the items A and B, where A beats B.) On the diagram that contains only counterclockwise arrows, find two existing neighboring items — we'll call them X and Y — where A beats Y but where A and Y both lose to X. For now, don't make any other decisions about which items A or B can beat.

Insert A between X and Y. Insert B diametrically opposite X. Now redraw the ring, spacing the old and new items evenly. Draw counterclockwise arrows between every pair of items, as before. Finally (and this is the hard part), come up with interpretations for all the new interactions, as represented by the arrows pointing to and from A and B.

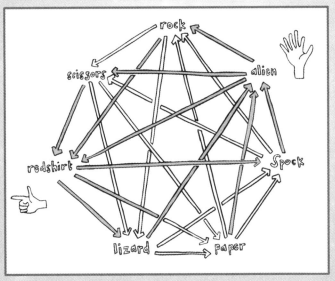

Look — we've added Yeoman Redshirt . . . and an alien!

The enhanced game above has two new items (it's now rock-paper-scissors-Spock-Redshirt-alien-lizard) and eleven new interactions, indicated by the dark arrows. We know that alien ambushes Redshirt (every darn time), but what do the other ten arrows symbolize? Let's see: Redshirt treads on lizard; Redshirt runs with scissors; Redshirt gets more battle-bikini-clad space babes than Spock . . .

---

*   Along with rock-paper-scissors and rock-Spock-lizard, there are three other balanced three-gesture games inside rock-paper-scissors-lizard-Spock. Can you find them?

EUREKA! @ CALTECH.EDU

"Be Serious — We're Playing a Game Here"[1]

Economist and political scientist Thomas Palfrey is the director of Caltech's Social Science Experimental Laboratory, a place where games are played in dead earnest. Unlike games played for fun, these activities are designed to give researchers insight into how people make decisions. In a typical experiment, groups of volunteer participants are invited to cooperate or compete on a computer-based task but are given only limited information about one another's actions. The computer collects each individual's decisions and determines the outcome, which may involve actual cash payouts or other incentives. The results are of great interest to those studying economics, voting, and neurobiology, among others.

For example, Palfrey recently studied a class of situations called allocation games, in which some number of participants must divide that same number of nonidentical items among themselves. In real life, this scenario occurs commonly, such as when assigning positions for a pickup soccer game, chores to household members, or roommates in a housing lottery. In Palfrey's experiments, participants requested specific amounts of money. There were as many predetermined amounts to choose from as there were participants. If everyone requested a different amount, then each participant received the amount he had requested. But if two people requested the same amount, then no one received anything. As the group tried again and again to find and maintain a satisfactory allocation strategy, with varying restrictions placed on their ability to communicate with one another, scientists were able to study such factors as fairness, efficiency, and patience.

In What Universe?
The Folded Building

At first glance Leonard, Sheldon, and Penny seem to live in an exceptionally ordinary apartment building: exposed-brick stairwell, broken elevator, cumbersome fire doors, basement laundry facility. But gradually it becomes apparent that the building is far from ordinary — and not just because its main door swings both ways yet somehow locks with a key.[2]

For one thing, where apartment 4A has only a tiny closet off the living room, apartment 5A (directly above it) has an entire bedroom. Moreover, the building is at least six stories tall (we can see the "up" button beside the fifth-floor elevator door), a zoning violation in most residential areas of Pasadena. In fact, the sixteen mailboxes in the lobby suggest that there are more than six floors, since as seen from the sidewalk, the building is far too narrow for more than two apartments per floor.[3]

Regardless of the number of stories, in what other building could a pizza delivery guy cover the three flights from the first floor to the fourth by walking up "like, four flights of stairs"? Evidently the extra flight works both ways: Penny's apartment is four stories above the basement laundry room, yet she gets from the one to the other by going "down five flights of stairs." And speaking of the basement, what are we to make of its miraculous window, which looks out on broad daylight from a position at least six feet underground?[4]

Fans have gone to great lengths to attempt to map out the internal structure of the building.* They've had to make allowances for portions we never see, such as any of the ceilings and the apartment of the "people on the first floor moving out" who sold the couch to Leonard.[5] What they ultimately find is that the building contains some strange twists and turns and eventually folds back in on itself, along the lines of the hypercube-shaped residence in Robert A. Heinlein's science-fiction classic "And He Built a Crooked House." Unlike the bizarre building in that short story, however, this one's footprint isn't even remotely rectilinear. The brickwork glimpsed through the kitchen window, for instance, appears to be the outside wall of Leonard's bedroom. You might like to try sketching out a rough floor plan for yourself.

Yeah, good luck with that.

---

*   An exact replica, in disconnected pieces, has been constructed on a sound stage at the Warner Bros. lot in Burbank. The stairwell consists of a single stairs-and-landing combination that extends down into a pit jackhammered into the floor. The stairs leading up make a ninety-degree turn at the end and continue on to a backstage area. A second set represents the building lobby.

---

1. "The Zazzy Substitution" (Season 4, Episode 3)
2. "The Misinterpretation Agitation" (Season 8, Episode 7)
3. "The Dead Hooker Juxtaposition" (Season 2, Episode 19) and "The Itchy Brain Simulation" (Season 7, Episode 8), respectively.
4. "The Deception Verification" (Season 7, Episode 2) and "The Speckerman Recurrence" (Season 5, Episode 11), respectively.
5. "The Staircase Implementation" (Season 3, Episode 22)

TWENTY-FOUR
COMING TO THINK OF IT

Serves Sheldon right for trying to turn a mother's relief at her boy's safety into a lecture on deductive reasoning. His aim is laudable — to explain to his mother something she understands only intuitively (if not Inuitively) — but his timing's rotten.

Deductive reasoning is a tool for taking one or more statements stipulated to be true and using a well-defined set of rules to deduce additional statements that must also be true. It's our best line of defense against logical fallacies (see "Nuh *Uh!*"). For example, if we are told that it is true that all men are mortal (meaning that no one lives forever) and that Socrates is a man, then we can conclude that Socrates is mortal — in fact, we can prove it.

post hoc ergo propter hoc Latin for *after this, therefore because of this.* This is the name of a logical fallacy in which the mere fact that one event happens to follow another is taken as evidence of a cause-and-effect relationship. For example: "Those telemarketers always wait until you're in the bathtub to call you." Which, if it really were true, would just be so incredibly creepy.

You might not need a rigorous proof in order to conclude that anybody, whether his name is Socrates or anything else, is mortal. It seems fairly intuitive. And if you know that this particular Socrates died over 2,000 years ago, that would support the claim that he is (make that was) mortal. But that's just hindsight talking. Intuition is debatable, and hindsight is no way to make predictions. That's where deductive reasoning comes in. (Woody Allen sent up deductive reasoning in the film *Love and Death*, when his character declares that "A. Socrates is a man" and "B. All men are mortal," and therefore "C. All men are Socrates."[1])

To Be or Not to Be (Socratic Version)

Let's prove that if all men are mortal, and if Socrates is a man, then Socrates is mortal. We'll use the technique of reductio ad absurdum (see "Nuh *Uh!*").

First, we assume the contrary of what we're trying to prove: that Socrates is not mortal but immortal. Well, if he's immortal he can't be a man, since all men are mortal. But we're told he is a man, so he both isn't and is a man. (There's the absurdum. We're done.)

Alternatively, we could have said that since the immortal Mr. S. is a man, we know of at least one man who isn't mortal. Ah, but we're told all men are mortal, so none of them is immortal and yet at least one of them is. (Absurdum again.)

No matter how we slice it, our temporary assumption ("Socrates is immortal") causes a logical conflict. Unless we decide that something else in the syllogism is not true,* we must conclude that our assumption was false and that Socrates, alas, is mortal. Sorry, Socs.

---

* For instance, *are* all men mortal? Throughout history it's been claimed that various personalities were immortal or were both mortal and immortal. Fortunately for this illustration (though perhaps less so for him), Socrates wasn't one of them.

This particular logical argument is fairly easy to navigate, but logical questions can get pretty hairy pretty fast. If all real ravens are black and Poe's raven was imaginary, then what color was it? Is the statement "This is a lie" a lie? If there's no business like show business and if show business is like show business, is show business a business or isn't it?

Deductive reasoning isn't just an exercise in head-scratching. Rigorous proofs are crucial because (as we've found) life just runs so much more smoothly when we know what's true and what's not and when we can predict the future. The simplest proofs are no less important than the most complex ones: they all come together to build the same rock-solid foundation.

> **argument (or syllogism)** A family of logical statements, consisting of premises (e.g., all men are mortal + Socrates is a man) and a conclusion (e.g., Socrates is mortal). If by the rules of logic the premises fully imply the conclusion, then the argument is *valid* — even if the premises are complete hogwash. If the argument is valid and in addition the premises are stipulated to be true, the argument is *sound*.

Interestingly, now that we've proven that the outlook for poor old Socrates isn't so rosy, it's worth noting that the foundation we used *isn't* rock solid. We assumed that the premises were true, but were they? It might not be the case that all men are mortal. You may strongly suspect it to be true, but it's not impossible that someone currently living, or someone who hasn't been born yet, will be immortal. (As it happens, there's no way either to prove or to disprove that: you'd have to wait until someone had lived forever, which would take a long time, or until there was nobody left, including you; see "Past Performance Is No Guarantee.")

For that matter, it's possible that "Socrates is a man" isn't true either, whether you take the word "man" to mean "human" or in any of its alternate senses, such as "adult male" or "mensch." Socrates could have been a wrinkled, gray-haired teenaged girl stuck in a time warp, or a milquetoast robot alien. (There aren't a lot of historians who will back up these sorts of claims, but strictly speaking they're possible.) For that matter, he could have been a zombie. So although we might disagree about whether the argument is sound, at least we know it's valid, because we've taken certain statements stipulated to be true and applied the rules of logic to them to prove the truth of an additional statement.

Because we're able to evaluate not only the validity of arguments but also their soundness, deductive reasoning permits us to build a structure of proven statements on a foundation of other statements, after which those new statements become part of the foundation for further statements. And if you're wondering what's at the bottom of that foundation, and how we can prove the fundamental premises that are then used as the basis for proving everything else — we can't. We simply have no way to "know" the nature of reality; the closest we can get to it is our own individual view of it. There's plenty of overlap and agreement between views, but at bottom each of us has to base what we call Truth on faith (see "KISS and

Tell"). How do we know there's any objective reality at all? Perhaps our collective views — which are all we have — are all there is. This cheerfully egocentric school of thought (you *are* the center of the Universe — your universe, anyway) is called anti-realism.

What "Is" Is

When you find yourself discussing what reality actually "is," you're peering into philosophy, the historical foundation of science. What then is the foundation of the foundation? Can we ever truly perceive reality, or are we doomed to gaze only at its reflections and projections, like one of Plato's after-dinner shadow-puppet shows?

And just as there's no art without the viewer, no time before the Big Bang, no temperature below absolute zero, no space beyond the edges of the Universe, no definitely-alive-or-dead cat until Schrödinger opens the box (see "Four and a Half Lives"), would there even be any reality without things and events to fill it, without at least one consciousness to perceive it?

"I can't be impossible: I exist," declares Sheldon with conviction, but isn't that really just wishful thinking?[2]

Still, there's something glumly unsatisfying about a worldview in which everything boils down to a matter of opinion. For one thing, how can you prove you exist? The French philosopher René Descartes tried with "I think, therefore I am." But what exactly does "I think" mean? What does "I am" signify? Who is this "I" person, anyway? Unless we can come up with a concrete understanding of those terms and their implications, something that doesn't lead us in circles and doesn't rely on anyone's opinion, the statement that there's an "I" that can "think" doesn't unequivocally prove that it "am."

To be fair to Descartes, he may have been saying only that in his opinion, the fact that he could do something he thought of as "thinking" constituted sufficient proof that he was doing something he thought of as "existing." Great conversation starter, René — and now you *don't* even exist anymore. (That's what comes of putting the hearse before Descartes.)

It seems to come down to this:

  1. I can't think unless I exist.
  2. I can think.
  3. Therefore, I exist.

On the one hand, this is perfectly valid deductive reasoning. As long as you assume that the first two statements are both true, then the third statement *must* be true; there's no way it could be false. But is the reasoning sound? We don't have much proof that those first two statements *are* true. Perhaps there are some things that can think but don't actually exist. (Your imaginary friends, for instance. Or maybe Descartes, although he no longer exists here, is somewhere else and still thinking things over. With Socrates, probably.) Or it might be that what we're doing when we *think* we're thinking isn't really thinking. Perhaps what we think "exist" means isn't actually right. The Universe could have been created last Thursday, carefully made up to look ancient and with all of us preprogrammed with a set of what we think are our memories extending back years. We could all be brains floating in a vat of nutrients and wired into a supercomputer that feeds us false perceptions (see "KISS and Tell").

Maybe we're one single consciousness that refuses to recognize itself.

Worth thinking about, perhaps. Hope it doesn't keep you up all night.

EUREKA! @ CALTECH.EDU
The Arrow of Time

When we talk about "cause and effect," we're using the rules of logic to tie together events separated in time. That's pretty bold, considering we don't even know what "time" is. (Other than "the thing that passes." Which it does at a speed of one second per second. Which isn't a very helpful definition.)

In general, the laws of physics are time-agnostic: a movie of two particles (or two billiard balls) colliding looks just as convincing when you run it in reverse. If the movie shows sixteen billiard balls coming together and fifteen of them come to a dead stop in a perfect triangle, your credulity might be strained. But even that is an utterly valid physical scenario — just very, very unlikely.

Caltech theoretical physicist Sean Carroll, the webmaster of PreposterousUniverse.com, is intrigued by the arrow of time. Why do we perceive the past and the future differently? The Second Law of Thermodynamics says that entropy (a measure of disorder or randomness) always increases as time passes. Scrambled eggs don't unscramble themselves, spilled milk doesn't unspill, and a bad smell that starts near you will eventually spread throughout the room. Entropy may be time's signpost, in that what we perceive as the future may be nothing more than the direction of increasing Universal entropy. We may be characters in a movie that's already been fully mapped out, aware of nothing but our present state — portions of which (our memories) can be influenced by events occurring in the lower-entropy past but not the higher-entropy future.

Or perhaps the arrow of time is simply defined as the direction in which effects are preceded by their causes. But that doesn't explain how the movie starts (see "KISS and Tell"). It doesn't say what minimum-entropy miracle put our whole Universe into its initial hot, dense state. Did an unimaginably vast wad of ultra-compacted material just happen to spring into existence in a fit of improbability? Did the Cosmic Egg simply drop unannounced from some Great Primordial Chicken?

For that matter, how can physical laws that work the same in both directions presume to distinguish between causes and effects? It's easy to conclude that the future never influences the past, but just as anything can be turned into a conspiracy theory simply by relabeling cause and effect (so that the later event's inevitability becomes the thing that makes the earlier event necessary), is it somehow possible for events that haven't happened yet to arrange things in a way that ensures their own inevitability? Maybe every causal connection we've ever made is only a post hoc ergo propter hoc logical fallacy. Do we exist as a result of the Big Bang? Or did the Big Bang happen so that one day we would exist?

"Imagination and science together is what enables creativity."

Special effects master Ray Feeney (Caltech '75) is the founder of two companies that provide leading-edge visual effects solutions for feature films, television shows, and commercials. A founding member of the Visual Effects Society, he helped to develop one of the first motion-control camera systems early in his career. His advances in motion-control systems, film recorders and scanners, and blue-screen chroma-keying technology have won him multiple Academy Awards for Scientific and Engineering Achievement.

Only one scientific or technical Academy Award — the Gordon E. Sawyer Award for "technological contributions [that] have brought credit to the [motion-picture] industry" — comes with an actual Oscar statuette. Feeney has that one, too.

---

1. *Love and Death*, directed by Woody Allen (Beverly Hills: United Artists, 1975), film.
2. "The Gothowitz Deviation" (Season 3, Episode 3)

TWENTY-FIVE
IT'S A FUZZY OLD WORLD

> Sheldon: I'm effectively one of Heisenberg's particles. I know where I am or I know how fast I'm going, but I can't know both.
> — "The Monopolar Expedition" (Season 2, Episode 23)

Cue polite applause. (Hey, if you weren't willing to tolerate the occasional esoteric science joke, you could've watched something else. *Breaking Bad,* perhaps? There's a Heisenberg in that one too, but the science humor is a bit darker.)

Sheldon's referring to the uncertainty principle formulated by physicist Werner Heisenberg. It places a fundamental limitation on how precisely we can measure certain things simultaneously.

Rulers, clocks, bathroom scales, electron microscopes: all measuring devices are imperfect. There's always some fuzziness in their results, some plus-or-minus factor to acknowledge that a follow-up measurement under identical conditions could return a slightly different result. Even with continued improvements in equipment and techniques, every physical measurement is understood to imply not merely the one value that happens to be reported but a range of possible values. These uncertainties frequently go unacknowledged — in the elevation and population numbers on the signposts at city limits, for example — but they're always there. And at the subatomic level, the problem gets compounded.

Suppose you have a particle (say, an electron) and you want to know something as basic as where it is and where it's going. To do this, you would need to measure two things: the particle's momentum (its mass, speed, and direction) and its position. But particles present a challenge to our intuitive understanding of the permanence of objects, because unless they're actively interacting with something else, certain of their properties are not merely unknown but undefined.

Worse, during subsequent interactions those properties may take on new values that can't reliably be predicted (see "Says You!"). Therefore, in order to measure either the position or the momentum we'll need to interact with the particle, and if we want our results to be self-consistent, we'll need to measure them both at once. (If we were to measure first one property and then the other, both properties could change between interactions, making the information gleaned from the first interaction obsolete.)

Building a system to measure a particle's position and momentum simultaneously isn't difficult, and there are ways to minimize the uncertainties in both values. But strangely, it turns out that beyond a certain point, anything we do to our system to lower the uncertainty in either one of the properties will cause the uncertainty in the other property to go *up*. No matter how cleverly we construct and operate our measuring device, the more accurately we make it pinpoint the particle's position, the less we can trust the momentum it reports, and vice versa. This unavoidable trade-off of fuzzinesses is a consequence of the graininess of reality (see "The Gravity Situation"), and there's simply no way for any physical device to avoid it.

The same trade-off happens with certain other pairs of observable properties — for example, a particle's lifetime and its energy. Inside an accelerator, we might like to make simultaneous measurements of both of these quantities as particles pop into and out of existence. But the more accurately we measure the one, the less faith we can put in our measurement of the other.

And that's what Sheldon is grousing about.

Many people use the label "uncertainty principle" for what is actually the observer effect. Heisenberg himself sometimes blurred the lines between the two — possibly revealing his own uncertainty about uncertainty.

> **observer effect** The impossibility of observing anything without affecting it in some way. The observer is also the observed; the journalist becomes part of the story. It's a corollary of the fact that there ain't no such thing as a free lunch.

The observer effect recognizes that you can't measure anything without having some effect, no matter how tiny, on the thing being measured. Shel Silverstein's poem "Stone Telling" recommends one effective method for finding out whether a window is open: just throw a stone at it. (If you hear glass breaking, then you know the window was closed. The only problem is that whether or not it was closed before, it isn't closed now![1])

The uncertainty principle is a fundamental limitation, and nothing can be done to get around it; it's just the way the Universe is made. By contrast, the observer effect can often be mitigated with cleverness: throwing a smaller stone at that window, for instance, or tossing it more gently.

An electronic version of Silverstein's window is found in the memory banks of early computers. They used magnetic cores: arrays of tiny iron doughnuts, any one of which could be magnetized by applying a small electric current. The magnetization, which persisted after the power had been removed, could be made to point in either of two directions, thus representing a single bit of information. To read back the bit later, another current was applied. Unfortunately, each bit could only be read once, because reading a doughnut had the side effect of erasing it. Any electric current weak enough not to disrupt the direction of magnetization was also too weak to read it reliably.*

----

* Minicomputer pioneer An Wang racked his brains for a long time before hitting upon the (tedious, but effective) solution

Here's an example of how the observer effect is tied to physics. It's impossible to spy on a particle unnoticed. There's just no way to observe one without affecting it. It's like trying to find out what a spinning coin feels like: the moment you touch the coin, it stops spinning. Or like trying to find a moth in a dark room by waving a broom around: until you feel the broom contact the moth, you only know where it isn't, and the moment you do happen to contact it, you'll send it sprawling in a random direction, and then you can't say where it is, only where it was. You could try using a floppy broom made of baby's hair so as to give the moth less of a shove, but you'll still probably never be able to touch it gently enough to avoid moving it. And the bristles of that floppier broom will be more flexible, which will make it harder to determine the exact moment of contact.

So the trade-off is: feel almost exactly where the moth was-and-now-definitely-isn't by whacking it hard (and giving it a tremendous random velocity), or probe very gently and change its velocity only a little (but be less certain about where it was).

It might be nice to have an idea about where you might find the moth the next time you look. One way to approach this is to divide the space into different areas and talk about the probability of finding the moth in each area. The higher the probability of finding the

---

to the demagnetization problem: after the doughnut's state is read out and thus destroyed, it's simply written back in again. (You were just about to suggest that, weren't you?)

moth in a given area, the more likely we'll observe it there. For example, you might decide that there's a higher probability of finding it near the center of the room than in the corners, and that there's a higher probability still of finding it fairly close to where you last found it, but a low probability of finding it in the exact spot where you last found it, because you remember whacking it pretty hard.

Probabilities can change with time and with new information. And they're only probabilities — the likelihood of an event — but they can be useful guideposts. However, you'll still have the problem of never being able to say where the moth *is*, only where it *was* when last observed, because you just can't touch it gently enough not to disturb it in some way.

It's the same problem at the subatomic scale. Any external interaction, no matter how gentle, will disturb a particle, which is a pity, since it means we can't observe all of its properties without altering some of them. We might, for example, like to track the position of the lone electron in an ordinary hydrogen atom. We can see where it is right now — say, by bouncing a bit of light off it — and get a fuzzy answer ("when I last checked, it was somewhere over near the left-hand side of the atom"), but we can never make that fuzziness go away completely. If we poke at it sharply, with high-energy light, we'll have a better idea of where it *was* at the moment we contacted it, but we'll knock it for a loop and have less of an idea of where it *went*. If we use low-energy light (the equivalent of our baby's-hair broom), we'll disturb it less, but we'll also be less sure of where we contacted it.

An electron doesn't orbit its atomic nucleus like a little planet; it smears out into a cloud of probability (see "Slits and Stones"). We can assign different probabilities to different regions of space around the atom, just as we did with the moth, and we can compute the probability of finding the electron in each region.

However, electrons in atoms have an additional oddity, which is that they don't travel from place to place; they only appear. With a moth, you know that at every moment it's always somewhere, and you know that it travels from one point to another by passing through a chain of points in between. An electron doesn't travel along a path like a moth. Observing an electron at point A and then later at point B doesn't mean it followed a path from A to B. It only means it was at A and then later it was at B. You can't be sure where you'll find it the next time you look or where you would have found it if you'd looked sooner.

This can be hard to picture. It might help to consider reality not as some absolute collection of objects "out there" but as "the collection of things that affect your personal universe" (see "Coming to Think of It"). Things that don't affect your universe aren't part of your reality. If something's affecting you in any way (even if you haven't actually observed it or are unable to observe it), it's part of your reality. If something isn't affecting you directly but it's affecting something that's affecting you, it's part of your reality. But if something isn't affecting you and isn't affecting anything that's affecting anything that's affecting . . . (etc.) anything that's affecting you, then it's *not* part of your reality.

This seems pedantic, but it hides a subtle complexity. If nothing is interacting with the electron in your hydrogen atom, so that nothing in your universe has changed as a result of that electron's position, then the electron's position is not part of your reality. The electron is there — it's part of the atom — but its exact position isn't part of your reality.

Unknown vis-à-vis Unknowable

Not knowing something because *you* don't know it yet is different from not knowing something because the Universe doesn't know it. When you're about to open the door of a wardrobe in a strange house, you may well feel that anything is possible and that it could as easily lead to a tunnel (or a brick wall or a snowy forest or the garden of the Queen of Hearts) as to a wardrobe.

But the Universe knows. When the wardrobe was assembled, the Universe changed in ways that depended on what was behind its door. What's beyond that door is already a part of your reality, even if it's not a part of your personal knowledge yet.

As with the moth, we don't have a way of knowing its current position, only its most recently observed position. But since electrons in an atom don't follow traceable routes like moths do, it can be observed at one point and then another without passing through a chain of points in between. Anytime something interacts with that electron, something about the Universe might change and make the electron's position become a part of your reality. But in the absence of any explicit interaction, there's no difference between a universe where the electron is *here* and a universe where the electron is *there,* and it's meaningless to say that it has any position at all.

It's not that it's somewhere and we just don't know where yet because we haven't checked. It's not that it's in some in-between position. It's that unless its position makes a difference, no matter how small, to *something* in the Universe, unless there's a way to distinguish between the two states "it's over here" and "it's not over here," it simply has no specific position. It only has a certain probability of being observed in each of its possible positions.

state A description of things,
not necessarily as they are, but as they might be.
superposition The overlap of two or more states, each with its own probability of being observed.

All the states that can describe a physical system can be thought of as overlapping one another, not necessarily in physical space, but more the way two different songs can overlap each other in your head. That's called a superposition of states, and it just means a collection of states, each with a probability of being observed. Conditions may cause the states and their probabilities to shift, sometimes dramatically. When we finally observe the system, what we'll see is one of the states whose probability at that moment is non-zero — but we can't predict which one.**

In our daily lives we don't usually notice all of this subatomic indeterminacy, but what if you built a device to amplify the superposition of states to our human scale? You could take two possible outcomes of some unpredictable event and map them to a pair of events that are strikingly different from each other. That's what Schrödinger did with his famous thought experiment involving a cat (see "Four and a Half Lives"). It's hard to do in real life, but it can be done in principle.

Again, the superposition of states means that things that aren't interacting with you in any way, including indirectly, don't have a definite state from your point of view. It's not that there's an answer "out there" and you just haven't learned it yet — it's that as far as your reality is concerned, there's no answer.

We have to be careful about using words like *is* and *will* and *exists*. You can *reflect* that something has happened in the past, you can *expect* that it will turn out to be the case again, but it's important to remember that nothing *is* until it, or its effects, are observed — and a moment later, it's not *is* anymore; it's only *was*.

Your reality consists only of those events that are

** Technically, everything is always in a superposition of states except at the moment of observation (meaning interaction with your universe). It doesn't matter if it's been observed in the past; it's now back in a superposition of states again (possibly a different one) because it hasn't yet been (re-) observed. And one moment after it's been reobserved and that observation has moved into the past, it will be in a superposition of states again. Whoa.

affecting you (including events th
by virtue of their having happene
someone else's reality consists of th
affecting him, and it's a dead certa
realities don't exactly coincide. Not
view — your realities.

Why a Certain Amount of Uncertainty I

The reason there's always a trade-off betv
has to do with the wave/particle dualit
"Slits and Stones"). Any particle (say,
be represented by a wave of "particle-n
through space. The wave is a combin
frequencies (each of which represents o
for the particle's momentum), and its
indication of the particle's position (in th
more likely to be found near the highest an
the wave than where it's close to its middle

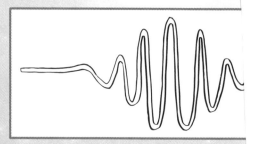

A slice through a model of the wave repres
particle. The horizontal direction represent
spatial dimension, and the vertical directio
"particle-ness."

This wave doesn't just sit around. Each time anything is interacting physically with the particle, such as when we make an observation, the wave's shape changes in a way that reflects the nature of the interaction. By their very nature, certain types of physical interactions are highly sensitive to the particle's momentum; during one of these interactions, the wave resembles an endless roller coaster, with the momentum corresponding to the distance between peaks. Other types of interactions are highly sensitive to particle position; these cause the wave to take on a narrow, spiky shape. Interactions that are moderately sensitive to both position and momentum cause the wave to look like a blend of the two: a roller coaster that fades out quickly. But no possible interaction is highly sensitive to both position and momentum — which makes sense, since there's no way one wave could look like a narrow spike and an endless roller coaster at the same time.

The wave is created by adding together many waves of different frequencies, each representing one possible value for the particle's momentum. The result is an interference pattern, with high crests and low troughs in regions where the waves happen to overlap constructively. Everywhere else, crests and troughs interfere with one another and the wave stays close to its middle value.

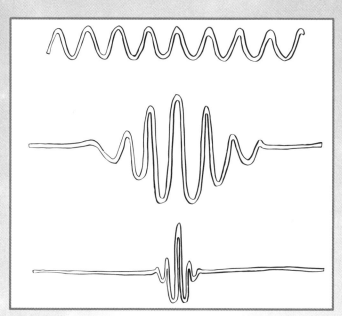

The particle waves from three different interactions. *Top:* Insensitive to position, highly sensitive to momentum. *Center:* Moderately sensitive to both position and momentum. *Bottom:* Highly sensitive to position, insensitive to momentum.

When an interaction allows for a wider range of momentum values, waves whose frequencies correspond to the additional possible values are incorporated, and the area where the particle is most likely to be found grows narrower. In this way, increasing the uncertainty in momentum decreases the uncertainty in position, and vice versa.

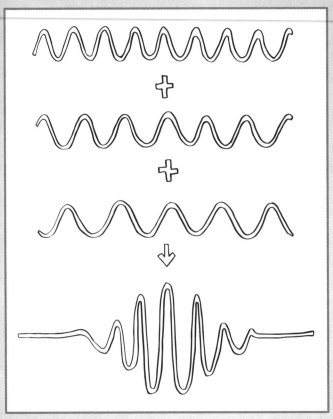

The wave results from adding together several waves of different frequencies, each representing one possible description of the particle at the moment of observation. (For simplicity, only three are shown.)

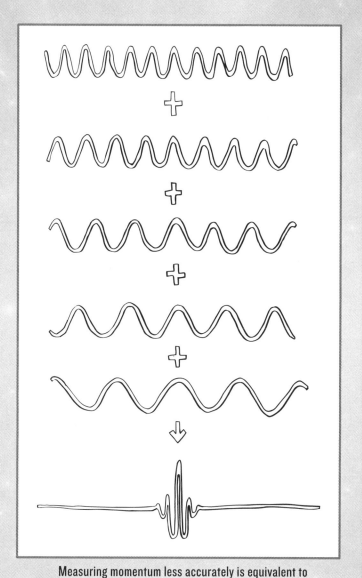

Measuring momentum less accurately is equivalent to adding more waves, each representing an additional possible value that falls within the (expanded) range of uncertainty. The wave of "particle-ness" grows narrower, confining the possible position to a smaller space.

It's something like what you might see on an X-ray of a stack of wire window screens whose mesh spacings differ only very slightly from one another. Almost every part of the image would look gray, because holes and wires would be overlapping in roughly equal numbers. But the narrow region where mostly holes (or mostly wires) lined up would show up in stark black and white.

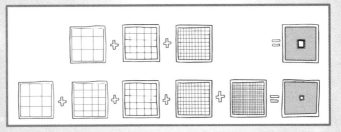

Window screens standing in for particle waves. (This visual analogy has a few shortcomings — wire we not surprised? — but it serves the purpose.)

As more screens are added, this region grows narrower. And that's why increasing the uncertainty in a particle's momentum decreases the uncertainty in its position, and vice versa.

Quantum computers — assuming a viable one can be constructed someday — would take advantage of atomic-scale indeterminacy to achieve enormous speedups over our present (classical) computers. Rather than processing only one set of inputs at a time, as classical computers must do, a quantum computer operates on all possible input combinations at once.

Suppose you're using trial and error to find out which numbers will divide evenly into a very large number. You could try dividing by one random value after another, hoping for early success, or you could use a quantum computer to perform all of those divisions at the same time.

In a classical computer, each bit of information can take on only one value at any moment: either a 0 or a 1. But a bit in a quantum computer (called a qubit, pronounced \CUE-bit\) can encode both a 0 and a 1 simultaneously. This is done by placing an ion, photon, or similar entity into a superposition of two states (representing 0 and 1) and keeping it carefully isolated from the environment, so that it can't be said not to be a 0 and it can't be said not to be a 1.

While the computation proceeds, the qubits that make up the quantum computer's memory can be thought of as representing all possible intermediate results. To read out the result at the end of the computation, they're simply observed, forcing each to become either a 0 or a 1. Nothing's certain, as Heisenberg says, but with clever programming and a carefully chosen initial state, chances are that the final readout will give the answer to the computation.

Repetitious, data-heavy operations that would take a

network of supercomputers billions of years to run (such as cracking modern data encryption systems) could theoretically take a quantum computer a few months or less. The only tricky part is figuring out how to build one.

[SCIENCE TO COME]
There Is No Know; There Is Only Nearly-Know

Heisenberg's principle is expressed mathematically by saying that the product of a pair of uncertainties can be no smaller than a certain very small (but non-zero) minimum amount. That means that neither uncertainty can be exactly zero; both of them must be somewhat greater than zero. (The only exception would be if the other uncertainty is infinite, but that's never the case for a real object, and certainly not for Sheldon.)

Sheldon implies (by not using a qualifier like "approximately") that he knows either *exactly* where he is or *exactly* how fast he's going. In other words, either his position or his velocity has an uncertainty of zero. But since his other uncertainty isn't infinite, the product of the two would be zero, which is smaller than the very small (but non-zero) minimum given by Heisenberg.

Sheldon's upset, he's bewildered, he's speaking off the cuff — and he does use the qualifier "effectively," so we'll go easy on him. But he's still wrong: he can never know exactly where he is, and he can never know exactly how fast he's going. Who among us can?

1. Shel Silverstein, "Stone Telling," in *Where the Sidewalk Ends* (New York: Harper Collins, 1974).

TWENTY-SIX
BUT THEN AGAIN, TOO FEW DIMENSION(S)

> Leonard: At least I didn't have to invent twenty-six dimensions just to make the math come out.
> Sheldon: I didn't invent them; they're there.
> Leonard: In what universe?
> Sheldon: In *all* of them — that is the *point*.
>
> — "Pilot" (Season I, Episode I)

Curious about what life's like in the fourth dimension and just not sure which way to turn your gaze? Don't feel bad. No one can visualize more than three dimensions. We can talk about higher dimensions and do math on them and describe some of the properties they would have to have, but no matter how hard we squint our mind's eye, we can't picture them. (In a deleted line from one episode of *The Big Bang Theory*, Sheldon scoffs, "Oh, please. I can visualize twenty-six dimensions. Don't you think I'm capable of manipulating a vehicle through three?"[1] Perhaps he can conceptualize twenty-six dimensions, but you can be certain he can't truly *visualize* them.) And that's perfectly normal. All our experience is in the familiar three-dimensional world, and it's very difficult for us to envision something we haven't experienced.

There's No "There" There

Science-fiction characters are always stumbling across the "entrances to" other dimensions — portals into hyperspace, doorways into alternate universes. But that's not the way it works. A dimension is a direction, not a location. You can't travel to or from it, only along it. In, not at. Whither, not where. Don't be misled by all the hyperspace hyperbole and portal prattle.

A dimension isn't a place; it's a direction, an extent, a basis for expressing the separation between things. With three spatial dimensions, we can measure the separation between points (e.g., one mile east and two miles north of here, three miles below ground). Throwing time into the mix lets us measure the separation between events (e.g., four seconds after sunrise). That's why time is sometimes called "the fourth dimension."

When Sheldon uses the Cartesian coordinates "(0, 0, 0, 0)" to refer to the moment he first sat on "his" spot, he's talking about the event whose position — "(0, 0, 0)" — and time — "(. . . , 0)" — he reckons everything from.[2] But this only works because he's sitting still. Literally. Einstein's theory of relativity is based on the observation that whenever two things are in motion relative to each other, space and time get tangled together. (Relative to each other . . . relativity . . . see what Einstein did there? And now you know why he called it that.) Instead of three dimensions of space (only) and one of time (only), we have to speak of four dimensions of space-time.

Space-time is harder to visualize than space or time separately, and to make matters worse, some models of the Universe require ten, eleven, or even twenty-six space-time dimensions. It doesn't matter that no one can visualize all those dimensions; the theories require them, or else the math doesn't work out. So if Sheldon wanted to give the full coordinates of the Historic First Sitting Event, he *should* have recited a lot more zeroes. (Let's be glad he didn't.)

But for now let's keep time separated from space — could everyone please just hold still for a moment? — and let's talk about a fourth spatial dimension. Not *the* fourth dimension — *a* fourth dimension. Three dimensions of space, one dimension of time, and now another dimension of space. Why? Because the physics of Nature strongly suggests that there are more dimensions than we're able to look in. (Yes, it's as distressing as when a child who has just gotten the hang of counting is told there are numbers below zero. How can there be less than none of something? How can there be a direction you can't look in?)

Consider a perfectly flat cinderblock wall with a dot painted on it. To give the coordinates of any point on the wall relative to that dot, we can measure the straight-line distance between the point and the dot and then note the orientation of that line. You could talk about the point that's three feet along a line that runs out from the dot at an angle of ten degrees above the horizontal. Those two pieces of information — distance and orientation — are enough to give the location of one single point unambiguously.

Another method, just as effective and possibly less wordy, is to take advantage of the regular grid made by the cinderblocks. You could say, for example, that the point is five blocks left of the dot and two blocks up. That combination of directions describes only one possible point; it's unambiguous.

In fact, as long as the cinderblocks are laid out in a regular grid, it doesn't matter whether they're parallel to the ground or even rectangular. You could just as unambiguously identify any point on a wall made of diamond-shaped cinderblocks or tipped up at one end.

Circles and Arrows

Both the distance-and-angle system and the distance-and-distance system are in widespread use. Which one you choose is a matter of convenience.

A Cartesian (grid) coordinate system — distance in each of two directions — works well for driving in the city. At sea or in the air, it's often more natural to use polar coordinates: range (distance away) plus bearing (angle around a giant imaginary circle).

There are also more exotic blends that combine the circles of polar coordinates with the distance-and-distance of the grid system. On the islands of Hawaii, it's common to give driving directions as distance along the (curving) shore plus distance inland or toward the ocean.

Everybody's Got an Angle

Manhattan is famous for its arrow-straight rectilinear grid, Tokyo for its ring roads. But Cartesian and polar grids aren't the only way of finding a position in two dimensions, as the slunchwise streets of Chicago demonstrate.

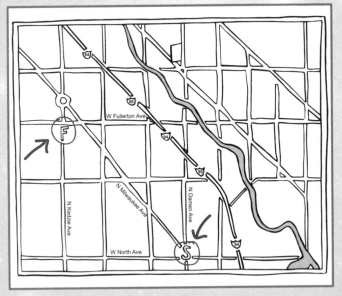

The asphalt in Chicago runs along three axes; any two will do

Most of the streets in Chicago run on a north-south-east-west grid reminiscent of the up-down-left-right grid found on maps and graph paper. You could use that grid to give the coordinates of Fullerton and Kedzie (the **F** on the map) relative to Damen and North (the **S**): two miles north (on Damen) and three miles west (on Fullerton). (These are coordinates, not necessarily travel directions, so it doesn't matter if they're roundabout or indirect.)

North-south is one dimension, and east-west is another. But the two dimensions you choose don't have to be perpendicular to each other. Any pair will do, as long as they don't point in exactly the same or exactly opposite directions. If you were restricted to only the diagonal roads and the east-west roads of Chicago, you could still give a perfectly valid set of coordinates: $2\frac{1}{2}$ miles northwest (on Milwaukee) and half a mile west (on Fullerton). Or if you were restricted to the diagonal roads and the north-south roads, you could say: three miles northwest (on Milwaukee) and half a mile south (on Kedzie). All three systems unambiguously give the coordinates of **F** relative to **S**.

What would be the coordinates of a point one inch in front of the cinderblock wall? Or a foot behind it? We could have the finest ruler in the world, but no amount of measuring *along* the surface of the wall could take us to either of those positions. To handle that third dimension, we must choose a third direction to place our ruler along: one that sticks out from the wall. Any direction will do, as long as it contributes something we can't get from the other two. It doesn't have to be exactly perpendicular to the other two; it just has to point at least partially away from the face of the wall.

By extension, if we wanted to add a fourth dimension to the coordinates, we would choose a fourth direction, not necessarily perpendicular to the other three but contributing something they don't. And here's where we run into difficulty. We don't know how to envision any direction that isn't just some combination of the length, width, and depth we're already familiar with. We have no experience of perceiving along a fourth, independent direction. Evidently it's not necessary for survival, because we don't seem to have evolved a way of doing it. So don't feel foolish or uninspired just because you can't imagine what the view is like along a new dimension; no one can.

It sometimes helps to imagine how you would explain a third dimension to a creature that only understood two. (Sometimes. For some people. Don't despair if you're not one of them.) For example, if you were in charge of highway development for the stick-figure people who live on the surface of a perfectly flat sheet of paper, you could draw lots of one-way signs for them. You could draw arrows pointing up the paper, down the paper, sideways; you could have long arrows for their highways and short arrows near their schools; you could have two-headed arrows for their two-lane roads and arrows looping back on themselves for their rotaries. Your two-dimensional friends would only be able to see those arrows edge-on, but they could move all around each one, feel its sides and corners, and understand which way it was pointing.

But how could you draw an arrow pointing even the tiniest bit off the surface of the paper? Could you make use of the laws of perspective, as scientists and chemists do, and draw the arrow wider at one end (the "closer to the viewer" end) or like a circled dot (a nose cone viewed from in front) or a cross (tail feathers viewed from behind)? Those cheats make sense to us because of the experience we already have in viewing things in three dimensions, but to the stick-figure people, they would just look and feel like very badly drawn arrows. They certainly wouldn't help anyone visualize the direction we call "above or below the sheet of paper."

This is what Edwin Abbott had in mind when he wrote *Flatland* (1884), the charming and thought-provoking book Sheldon mentions as one of his favorite places for an imaginary visit.[3] The two-dimensional people of Flatland inhabit a two-dimensional world, something like living drawings on a map. It's not just that they're *rather* flat, like a flat fish — they're *completely* flat. They have length, and they have width, but they have no depth. (We all know people like that.)

The denizens of Flatland are shaped like simple geometrical objects: squares and triangles and so forth, perpetually looking to the side like profile portraits. They can see one another edge-on, but what they see doesn't strike them as being flat, since they have no conception of thickness. A mysterious force that pulls slightly in the southern direction gives them a feeling akin to gravity.

Flatlanders move by sliding around on their flat world, unaware that a third dimension exists. Their two-dimensional universe completely fills their perception, and they can neither draw an arrow that points in any direction other than along the plane of Flatland nor peel themselves up off it.

When a spherical Being, cruising comfortably along in three dimensions, happens to cross through the plane of Flatland, the inhabitants experience it as a dot that suddenly and inexplicably appears, widens rapidly into a circle, and then shrinks to a dot again before disappearing. The sphere, hovering close by, sees the Flatlanders — their outsides and insides — as drawings on a page, and they can hear it speaking to them. But no matter how they spin they can never look toward it, and they're mystified when it speaks of a direction that is "Upward, and yet not Northward."

Though we mock the Flatlanders for their ignorance of what seems to us an obvious additional dimension, how must we look to a hypothetical being that can move through a fourth dimension? And why stop at four? If there really are twenty-six space-time dimensions, that's a lot of directions a spherical Being can stare at your insides from — that you can't stare back in.

But there's probably no need to worry about that. The electromagnetic field, which carries light, is most likely constrained to our familiar dimensions, so no one looking along any other dimension would be able to see you. In fact, gravity is the only field that's expected to leak out into other dimensions, and we haven't yet detected any such leakage. So if other dimensions do exist, they probably don't extend very far before wrapping back on themselves, which they can do because space can curve.

That's *Some* Diet

If an object of infinite thinness is too hard for you to envision, consider that Flatland could in fact extend into the third dimension. The Flatlanders, portrayed by Abbott as something like living geometric shapes sliding around on a tabletop, could just as well have been described as three-dimensional beings who are only aware of the parts of themselves (and others) that are in direct contact with the table.

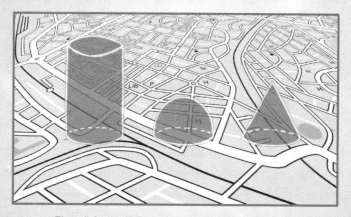

Three "identical" Flatland circles, out for a stroll

A cylinder, a hemisphere, and a cone could each have a circular "footprint" on the Flatland tabletop, so all three could appear as circles to one another and to anyone else in Flatland, but an observer capable of seeing into the third dimension could easily perceive their differences. In fact, despite their identical footprints, they could have very different masses — for reasons that would appear mysterious to them.

Going back to the notion of Flatland as a map, imagine that you could somehow bring the east and west edges of the map together. A Flatlander journeying east from his house would eventually, without changing direction, arrive at his own western door. He wouldn't feel the sudden jump from the eastern edge of the map to the western, because those edges are no longer edges; they're just as smooth as the rest of the map.

But surely he would notice the curve of the paper? Wouldn't he — especially if he were almost as wide as the map — recognize that Flatland was no longer flat? Not if there was nothing to indicate the curvature to him. He wouldn't see light curving around, nor feel any force bending him in one direction or the other, because light would follow the curve of the paper, as would gravity and the other forces. (Technically, it's the other way around: the paper follows the curve of the forces.) And if there's no experimental way of distinguishing between two things (even thought-experimentally), then the two things will appear to be identical. Just as with the superposition of states, it would be as meaningless to say that he feels the curvature of space as to say that he doesn't (see "It's a Fuzzy Old World").

But there are some observations that a clever Flatlander could make that would tell him his space was curved. If he can observe two distant objects and work out the dimensions of the huge triangle they make with him, he might discover that he can't draw that triangle on a flat piece of paper because one of its legs is too short or too long, implying that his space is curved. And if his eyes were sharp enough and Flatland were small enough, he could peer far to the east and see himself (or at least the back of his own head) as viewed from the west. Or he could look west and see himself as viewed from the east. It's not that he would see another Flatlander who looks just like him — it would be he, himself, *him*. If he could see even farther, he might be able to look beyond that *him* and see another *him* twice as far away. And three times. And four. Highly suspicious.

Suppose the northern and southern edges of the Flatland map were also joined, either to themselves (forming a globe) or to each other (forming a doughnut shape). Now our farsighted Flatlander can see additional views of the back of his own head by looking north, south, and in many other directions. If his arm were long enough, he could reach out and poke himself — any of the himselves — and at the same moment, he'd feel himself being poked. When one of him moved, they'd all move, because they're all the same *him*.

From his point of view, the curving universe isn't curved at all; it's an infinite flat sheet on which everything repeats exactly, at specific distances. The fact that from your "outside" vantage point in the third dimension you see the map curving (and the light and the arm) won't change his own perception. He can't *feel* the curvature of space, but (if he's a particularly clever Flatlander) he might recognize that this orchard of identical clever Flatlanders is being caused by it. The alternative — to tell himself that space doesn't curve and that he's simply living in a universe made up of lots of things that just happen to look and behave identically — fails according to Occam's razor (see "KISS and Tell").

Isn't that going about it the hard way? Can't he somehow "override" the curvature of space just by looking "straight" past it, the same way you can look at the eastern horizon and see the Sun, rather than Vladivostok? No, because that would mean somehow looking along a dimension that doesn't follow the surface of the map, and we know our Flatlander can't do that. Every light ray he sees comes to him along that surface, having followed its curves, and no matter how he turns, he can never break his vision free of it.

Regardless of whether a two-dimensional world wraps around along its surface, it could wrap around in a third dimension. Imagine that the space in front of and behind our Flatland map is so tightly curved that it extends only a tiny distance before wrapping around. The effect would be the same as if there were an infinite number of identical Flatlands, all stacked up like a pile of maps: the moment you rose above the front surface of Flatland, you'd be entering Flatland again from the back. A Flatlander who somehow developed a method of twisting his eye just the least bit away from the plane of the world would still be looking into Flatland, across an unnoticeably tiny gap. Rather than the whole new off-the-charts (literally) world he was hoping to find, he'd see essentially the same view.

That may be the situation we have with the higher-order dimensions Sheldon didn't invent. It's thought that at the time of the Big Bang, all the spatial dimensions were equally tiny; that is, their curvature was so small that a trip along any of them would return you to your starting place almost immediately. For reasons we don't understand, nearly fourteen billion years ago some of the dimensions ignored Ed Robertson's polite "Wait!" in the theme song, and a rapid expansion started. But all the other dimensions have remained tightly curved, and like the Flatlander looking up through the infinitely stacked Flatlands, we won't see much of a different view by looking along any one of them. When scientists say that higher-order dimensions are curled up tightly, they don't mean there's a tiny coil of something or other over *here,* and maybe a different one over *there.* They mean that anything that leaves the world along those dimensions will re-enter it less than an atom's width away. Not much of a shortcut.

And why did only some of the dimensions expand? That's a great question, and no one has the answer. But when Sheldon describes the process of removing his trousers through a fourth dimension (actually, he slips and says *the* fourth dimension, but that's just the alcohol talking), he's doubtless aware that it's probably impossible.[4] Not simply because we don't know how to direct a force in that direction, but because any additional dimensions are probably so tightly curved that there's no way to jockey a single atom through them without crashing it into itself — never mind a pair of atoms, much less a pair of pants.

Dude, Where's My Transverse Energy?

Sheldon: ¿Dónde está el boson de Higgs?
Leonard: En el acelerador de particulares.[5]

At the Large Hadron Collider, the world's largest particle accelerator (or, according to Leonard's halting Spanish, the "accelerator of particulars"), scientists slam atomic nuclei head-on into one another at nearly the speed of light and measure the energy of the debris that comes flying out sideways. Any energy that goes unaccounted for (called missing transverse energy) could be due to undetected particles but might also indicate energy leakage into other dimensions.

Caltech experimental physicist Maria Spiropulu sifts the rubble of these collisions for evidence that such extra dimensions exist. A member of the Large Hadron Collider collaboration that discovered the particle known as the Higgs boson, she's also vice-chair of the American Physical Society's Forum on International Physics.

Spiropulu doesn't understand Sheldon's disdain for experimental physicists like Leonard: "Without experimental physics, we could never have confirmed that the top quark exists. Or the Higgs boson. But often we stumble on discoveries we did not expect and hadn't designed the experimental apparatus for. The Hubble Space Telescope is a well-known example: it was built for imaging galaxies, but what it found was evidence of dark energy and the curvature of the Universe."

ASK AN ICON: Tom Lehrer

Before launching into his indecent four-dimensional proposal, Sheldon attempts to lead the audience in a rousing chorus of Tom Lehrer's "The Elements." Set to the tune of "I Am the Very Model of a Modern Major-General" from Gilbert and Sullivan's *The Pirates of Penzance*, the patter song cheerfully barrels through the names of all 102 of the then-known elements in the periodic table, in rhyme.*

In the half-century since it was written, several new elements have been named, but as the last two lines explain:

These are the only ones of which

the news has come to Ha'vard,

And there may be many others,

but they haven't been discavard.[6]

Lehrer wrote and performed satirical songs with fiendishly clever lyrics and always-perfect rhyme schemes in the 1950s and '60s. A longtime teacher of mathematics and musical theater, with a master's degree before the age of twenty, he is occasionally addressed incorrectly (but understandably) as "Dr. Lehrer."

Aside from rock-paper-scissors-lizard-Spock (see "Ick-Ack-Spock"), Lehrer's "The Elements" is the only *Big Bang Theory* material Jim Parsons has ever confessed to struggling with. Parsons termed trying to learn the song "literally one of the worst moments of my life."[7]

Q: Jim Parsons says now that he's committed "The Elements" to memory, it's "like a war wound — it will not go" away.[8] Can you offer him any consolation?

Tom Lehrer: It may be that Jim Parsons was simply born at the wrong time. A Sheldon living just a few millennia ago would surely have found Aristotle's version of "The Elements" [also by Lehrer] slightly less daunting:

> There's Earth
> And Air
> And Fire
> And Water.

Which, although it doesn't rhyme in Modern English, doesn't rhyme in Ancient Greek, either.

---

* "The Elements" has also appeared on an episode of *Gilmore Girls*, and it featured as a plot device on *NCIS* (or, as Penny renders it, "*NCSTD*"[9]).

1. "The Euclid Alternative" (Season 2, Episode 5) — but honestly, the line really was deleted.
2. "The Cushion Saturation" (Season 2, Episode 16)
3. "The Psychic Vortex" (Season 3, Episode 12)
4. "The Pants Alternative" (Season 3, Episode 18)
5. "The Re-Entry Minimization" (Season 6, Episode 4)
6. Tom Lehrer, "The Elements," in *Too Many Songs by Tom Lehrer* (New York: Pantheon, 1981).
7. Episode #341 (Dec. 5, 2012), *Conan* (Atlanta: TBS).
8. Ibid.
9. "The Hesitation Ramification" (Season 7, Episode 12)

TWENTY-SEVEN
RIGHT BACK ATCHA

Leonard: Hey, you know who'd really dig seeing this experiment? Penny.

Sheldon: I wasn't aware that lunar ranging was her thing. Although I suppose the retroreflector left on the Moon by Neil Armstrong does qualify as a shiny object.

— "The Lunar Excitation" (Season 3, Episode 23)

Okay, so evidently the astronauts left a mirror on the Moon, and with a bright enough laser you can bounce a beam of light off it on a clear Pasadena night. (This is sort of an extension of the childhood experiment of shining your flashlight straight up into the night sky in hopes of spotting — or being spotted by — an alien.) And since the beam of light travels at a known speed, if you carefully measure how long it takes to go out and back, you can compute how far away the Moon is.

But hold on a minute. Even if you could pinpoint the location of the mirror, what if your laser beam doesn't strike it head-on? It would get reflected at some crazy angle, and you'd never see it. (Though maybe those elusive aliens would.) No worries — someone's already thought of that. The mirror's a very special type: it always bounces your beam right back to you.

The next time you're at a gym or in a fancy bathroom, look around and try to spot a place where two mirrors meet at a right angle. When you look into the corner formed by those mirrors, what do you see? Yourself. No matter how close you move in, how far you back up, or whether you lean left or right or get behind one of the mirrors and peek around it from the back, right there in the corner you'll always see a reflection of yourself.

As with any flat mirror, the reflection appears to be just as far on the other side of the glass as you are in front of it. Unlike with a flat mirror, however, your reflection is not reversed left to right. This is easiest to recognize if you're wearing something with lettering on it or if you extend your right hand as if to shake hands with yourself. Your reflection extends its right hand too.

Mirrors that are mounted at ninety degrees to each other but that don't actually meet in a corner (such as the mirror over your sink and the one on the medicine cabinet on the adjacent wall) don't produce this effect, because the area where your reflection would be is mirror-free. (That's no reflection on you.) And the multiple mirrors used in clothing stores and tailor's shops usually won't work either, because they're set at a much wider angle than ninety degrees, so that you can see yourself from multiple angles.

In the '90s (appropriately enough), a company called True Mirror marketed a device constructed of two high-quality mirrors set at a permanent ninety-degree angle. The True Mirror showed the viewer as he or she appeared to others, rather than reversed as in a standard mirror. The vertical seam running down the center was made all but invisible by using first surface mirrors (mirrors with the reflective coating on the front of the piece of glass rather than on the back).

The True Mirror never really took off, perhaps in part because of its price point and fragility and the fact that it couldn't hang flat against a wall like a normal mirror, but most likely because it really threw people off. It doesn't matter that the face you see in the True Mirror is the one the rest of us are accustomed to seeing; to you it just looks wrong. It looks flipped the wrong way around, and that can be eerie and a little unsettling.*

You can create your own non-reversing mirror by holding a small mirror against a larger one and tilting it back and forth until you can see just one copy of yourself. If you close one eye, you'll see that what appears at the seam between the two mirrors is the

---

* In one episode of *The Big Bang Theory,* Leonard is in his car with the camera looking over his shoulder from the backseat.[1] Somehow, everything we can see in front of the car is also visible in his side-view mirror, unreversed — an example of neither a retroreflector nor a True Mirror but a rushed post-production supervisor.

reflection of the other eye.

Now rotate both mirrors off the vertical so that although they still meet at a ninety-degree angle, the seam between them is no longer running straight up and down. Your reflection appears to rotate off the vertical, in the same direction but twice as far. When the seam is at a forty-five-degree angle, your reflection is lying on its side. When the seam is running side to side, your reflection is standing on its head. But it remains an un-reversed reflection: you look (to yourself) the same way you would look to anyone else if you were standing on your head.

As long as both ends of the seam between the two mirrors remain equal distances away from you, you'll see yourself no matter how the mirrors are rotated. But tilt the mirrors away from you, so that one end of the seam is closer to you than the other, and you'll quickly fall (or rise) out of view.

By placing a third mirror perpendicular to the other two, you can keep yourself permanently in the picture. No matter what angle you view it from, you always see the same thing just at the point where the three mirrors meet: the reflection of your own eye — upside-down, reversed, but always visible. This configuration, an inside corner made of three mirrors, is called a retroreflector. A beam of light entering it from any angle will bounce off all three mirrors and be reflected back out in the exact opposite direction.

Retroreflective materials are used in reflective safety clothing and on highway signs, the kind that seem to glow overly brightly in the beam of your headlights. Tiny particles of glassy material in the paint or in the fabric capture incoming light and bounce it back out in the reverse direction. A retroreflective layer of tissue behind your cat's eyes is what makes them glow when it looks into a light. This layer may have evolved in order to double the amount of light falling on the retina. (Your cat could probably confirm this for you if she ever felt like giving you the time of day.)

There are five retroreflecting devices up on the Moon, placed there by three teams of Apollo astronauts and two unmanned Russian Lunokhod ("Moonwalker") missions. Each device is roughly the size of a pizza box and holds an array of retroreflectors. (Using an array rather than a single retroreflector multiplies the amount of light being sent back and also reduces the chance of failure.) All the arrays face toward the Earth, which hangs in an approximately unchanging position in the Moon's sky.[**]

A lunar retroreflector is made not by gluing three mirrors together, which could easily fill with debris or be knocked out of alignment, but from a solid triangular pyramid of fused silica, a pure form of glass with a high internal reflectance. Incoming light entering the base of this "corner cube" retroreflector reflects off the insides of its polished faces.

The original lunar laser ranging retroreflector

[**] Very much like a satellite in geosynchronous orbit (see "Hexagon with the Wind"), though not for the same reason. Since the Moon always keeps the same face toward Earth (not counting a few small wiggles), there's no Earthrise or Earthset to speak of.

experiment sits close to the center of the Moon's disk as viewed from Earth. It was placed by Apollo 11's Neil Armstrong and Buzz Aldrin in 1969, specifically to establish the Moon's precise distance from Earth. Those first measurements were accurate to within about ten feet. Over the intervening decades, improvements in electronics and optics (at this end of the experiment) have made it possible for the same device (at that end) to give answers that are accurate to within one inch, or roughly the height of one corner cube retroreflector. Since the overall distance is a hundred times the width of the United States, that kind of precision would be like measuring the distance from sea to shining sea to the nearest 100th of an inch.

ranging Measuring how far to without having to travel to.

The principle is simple: shine a light at the retroreflector array, detect the returning beam, and multiply the elapsed time by the speed of light to get the round-trip distance. To make sure it's *your* beam of light you're seeing, and to help it stand out against the background of all the other light coming from the Moon, you choose a very specific color and take multiple measurements.

If only it were that simple. One of the many complications is relative motion. In the second and a half it takes a light beam to cross the one-way distance, the Moon will have carried the retroreflector array a mile farther along on its orbit. By the time the light returns, the detector, rotating with the Earth, will have slid half a mile or so to the east. In fact, the surfaces of both the Earth and the Moon are constantly twisting, turning, bouncing, and sliding in a complicated three-dimensional dance. One of the factors at work at both ends is the solid-body equivalent of the ocean tides. The Moon's gravitational pull actually distorts the solid Earth, and vice versa. Additional terrestrial complications include the movement of glaciers (allowing the ground to spring up as they recede), the precession of the equinoxes (the wobble in the Earth's spin), and continental drift.

And along with all of these relative-motion corrections, there are problems of geometry and metrology. Since neither the Earth nor the Moon is perfectly round nor perfectly uniform, complicated math is needed to convert the surface-to-surface distance to a center-to-center distance, all the while taking into account that according to general relativity, light slows down in a gravitational field.

Then there's that same problem you had with your alien-signaling flashlight: spreading. Over moderate distances, laser light stays tightly focused (see "Laser Fair"), but no laser beam is perfectly collimated, and we're talking about a quarter-million-mile one-way journey here. Add to that the distorting effects of the Earth's atmosphere, and by the time the light reaches the reflector it will have fanned out into a circle a few miles wide. Though this helps solve the aiming problem, it causes a terrific dimming problem.

How Far Is *What* from *What*, Exactly?

It's an easy question to state but a hard one to define: how far away is the Moon? No doubt we'd all prefer a single, simple answer, along the lines of "How far is Minneapolis from St. Paul?" or "How far is your hand from your elbow?" On the other hand, if you want to get nitpicky, neither of those questions has a single, simple answer either. Which part of your hand? Which part of your elbow? Which part of St. Paul? What if you bend your wrist?

Lunar ranging is plagued with the same sorts of ambiguities. The Moon is receding from Earth and doesn't orbit in a perfect circle, so the distance isn't a fixed number. For that matter, some corner cubes will be nearer to us than others, and we have no way of knowing which one any particular photon happened to bounce off.

Still, the experiment's worth doing. For one thing, the huge masses and distances involved make this the best and largest laboratory ever constructed for testing out and refining the laws of gravitation. And gravitation is tied into relativity, and relativity is tied into *everything*.

The reflected light spreads out again on the return journey, so much so that even if Leonard uses the brightest laser available, and even if he shines it at the Moon continuously, he'll be lucky if his detector catches more than one returning photon every few seconds. (By comparison, the dimmest stars you can make out with the naked eye are tossing a couple of hundred photons your way each second.) The others will be scattered all over the Sea of Tranquility and Los Angeles County or will have taken a wrong turn somewhere in between.

But he wouldn't want to shine his laser continuously, because then when a photon did come in, he'd have no idea how long ago it had left Earth, so he wouldn't be able to compute its flight time and thus how far it had traveled, which after all is the whole point of the experiment. The solution is to send the laser light in short pulses (optionally using a telescope to focus the photons on their way out and to collect them on their return) and hope that at least one photon makes it all the way back to the detector. To get an accuracy measured in inches, each pulse has to be at most an inch long — a distance that light covers in a tenth of a nanosecond. And scientists have their sights set on shorter pulses still, in hopes of getting that number down into the range of millimeters.

Astronomy, nearly alone among the sciences, has labored for most of its history under an impaired capacity to perform experiments. Unable to reach out and manipulate the objects of their study, astronomers have largely been relegated to observer-only status. But now they have the biggest laboratory of all. Four decades ago, mankind finally dropped in on its cosmic neighbor and, to the delight of lunar geologists, brought back dirt older than dirt. As a housewarming gift, we left behind a few shiny things for anyone to bounce a laser off. Or a flashlight.

Lassoing the Moon

NASA's Jet Propulsion Laboratory, administered by Caltech, is home to a wide variety of space-related programs. And JPL is where Dr. Jim Williams does data analysis on lunar laser ranging (LLR) results.

The distance to the Moon, far from constant, is increasing an inch and a half every year, which means Williams is shooting at a wildly moving target. His analysis must take into account a variety of significant and sometimes surprising influences on the Earth-Moon system, such as Jupiter's gravitational tug, the physical dynamics of the Moon's liquid core, and relativity.

Discrepancies between prediction and observation constantly require him to tweak his already-complex models but have also led to new insights. For instance, certain variations in the distance to the Moon could be accounted for if it has a solid core inside a liquid one, as the Earth does.

The biggest help to LLR precision, says Williams, would be to build more lunar ranging stations, including some closer to the Equator. With multiple lines of sight, accuracy improves. "We are down to two regularly operating stations, which is minimal [for effective analysis]. If it gets down to one, that would be a big problem."

In the meantime, a Pasadena rooftop is a fine first step.

---

*   The two stations are Apache Point in New Mexico and Côte d'Azur in France.

OUT TO LANDS BEYOND

"I truly know what it is to be confused by science."

Sandra Tsing Loh (Caltech '83) is a writer, performer, and radio and TV commentator. The third Caltech graduate in her immediately family, Loh earned a B.S. in physics but to this day isn't exactly sure how; her diploma, she says, was made entirely of partial credit.

Immediately after graduation, to her father's horror, she fled the sciences for the world of performance art. She has conducted an orchestra on the Malibu sands during the twilight grunion spawning run and serenaded rush-hour traffic from a piano perched on a flatbed truck.

The creator of numerous radio series, books, and one-woman shows, she now hosts a daily radio broadcast targeting those who, like her, are fascinated by science but easily overwhelmed by its complexities. Each installment of *The Loh Down on Science* introduces a new development in science, couched in Loh's own brand of slightly bewildered humor.

In 2005, Loh became the first alumna to deliver a Caltech commencement address. Her message to the graduating class was: "Dare to disappoint your father."

---

1. "The Vacation Solution" (Season 5, Episode 16)

TWENTY-EIGHT
PAST PERFORMANCE IS NO GUARANTEE

> Howard: Look at Leonard's record: twenty-seven days with Joyce Kim . . . two booty calls with Leslie Winkle . . . and a three-hour dinner with Penny. . . . Based on the geometric progression, his relationship with Stephanie should have ended after twenty minutes.
>
> Sheldon: Yes, I'm aware of the math: y equals 27 days over 12 to the nth.
>
> — "The White Asparagus Triangulation" (Season 2, Episode 9)

How has Howard turned a scattershot list of rather pathetic assignations into a confident prediction? Why is Sheldon so quick to agree? And . . . "should"? In what sense "should" the relationship have ended? And after such a precise time period? And according to whom? Who's in charge of "should"? (Let's not even get into why a relationship that "should" have ended didn't. That's a glass house too many of us live in.)

The "should" in this case isn't a question of

morals or pessimism; it actually follows from a bit of math. It's an example of inductive reasoning, where observations of the past are used to make predictions about the future. Despite the similarity in their names, inductive and deductive reasoning (see "Coming to Think of It") differ in a number of important respects. The main one is that deductive reasoning uses facts to establish other facts, while inductive reasoning uses observations and assumptions to make predictions. Deductive reasoning, when done correctly, is unassailable; inductive reasoning, even if done right, is always assailable.

Deductive reasoning begins with a collection of statements stipulated to be true. Rigorously defined laws of logic are then applied to them, and out come one or more conclusions that must then also be true. The results are conclusive: as long as no laws of logic have been violated, it's impossible to reach a conclusion that's false if the premises are true.

The Luckiest Guesser in Fiction

Fictional consulting detective Sherlock Holmes, though much admired for his powers of deduction, is actually a master of *induction*. His declarations very often take the form: "I've frequently observed that X is true whenever Y happens, so since Y is happening now, it's likely that X is true."* He isn't saying that X *must* be true; he's making a shrewd guess based on his experience. He can rarely be 100% certain without additional hard evidence, such as catching someone in the act or receiving a letter saying,

---

"Right-o, Mr Holmes, you clever fellow! Indeed it was I, in the conservatory, with the lead pipe."

Suppose Holmes were to announce, "I deduce that this stranger with the anchor tattoo must be a sailor because all people with anchor tattoos are sailors." Though that would be a valid deduction, it wouldn't be a sound one. He may suspect that virtually all people with anchor tattoos are sailors — he may never in his life have observed a non-sailor with an anchor tattoo — but for all he knows, this very stranger is an exception. A more honest declaration would be something like, "Having observed a strong correlation between the displaying of an anchor tattoo and the being of a sailor, I *suspect* that this stranger with the anchor tattoo is a sailor." (To Conan Doyle's credit, Holmes often couches his deductions in this fashion, but Watson still goes all starry-eyed.)

In his pursuit of the truth, Holmes is often assisted by recherché knowledge, fortuitous coincidences, and bold assumptions (along the lines of "The trains are running late today, so the murderer must still be in town"), but as with every detective, real or fictional, his hunches can only offer tentative explanations for what he has observed. He just happens to be an extremely keen observer — and an even better huncher.

---

* Nothing in this section is an actual quote.

Inductive reasoning, by contrast, begins with a collection of observations and then applies pattern matching, hunches, and intuition to try to guess at a rule that accounts for those observations. Any conclusion can be disputed or negated at any time, in response to additional observations or fresh insights or different thinking.

Deduction leads to provable facts. Induction leads to plausible rules.

Induction is a reflection of the learning process and of the way we turn incoming information into knowledge. Every organism that has the capacity to learn does so by using some form of inductive reasoning, seeking patterns that can connect new phenomena to earlier experiences. This is what allows you to recognize that something you've never seen before is a flower or to draw conclusions (possibly terribly unfair ones) about people you don't even know (gee, it's high school all over again). It's also what gives rise to the aphorism "Good judgment comes from experience, but experience comes from bad judgment."

Most of the rules of Nature you carry around in your head came from induction. Sometimes all we need is a rule, any rule, that accounts for past observations. But rules are most useful when we can use them to predict and manage the future. Experiments are a way of testing those predictions. Hidden-rule games like Scissors, Eleusis, Gestalt Number Theory, and Mao require players to perform experiments and search for patterns in the results just to figure out the rules of the game. A desperate player can always hope someone will eventually just tell him the rules, but in real life, Nature never tells; it only drops hints.

Mini-Quiz: Answers
1. Deductive. We go from a statement about ravens in general to a conclusion about some specific ravens.
2. Inductive. We go from observations of some specific ravens to a guess about ravens in general.

Induction involves assumptions. You may have been told that when you assume, you make something out of u and me. But we make assumptions all the time. Language processing, optical illusions, magic tricks, and jokes rely on them. We often flag them with verbal indicators: "If all goes as planned"; "Hopefully"; "As far as I know"; "It seems to me"; "I suppose/guess/think." We also flag them for others, with or without their consent: "Well, technically"; "How can you be sure of that?"; "Says *you*!"

How unbearably tedious (and potentially dangerous) life would be if we could never assume! Imagine telling a friend, "I'll see you later," and hearing him respond, "You can't *know* that." The first few times it happened you could correct yourself ("Well, I *hope* to see you later"), but eventually you'd probably find yourself hoping *not* to see him later.

"Technically": The Game of Being Progressively More Annoying

Three logicians (lawyers/nerds/whatever) are riding in a train. As they enter Scotland, the first thing they see is a black sheep standing alone in a field. One comments, "I see that there are black sheep in Scotland."

The second pipes up, "Technically, all you see is that there is at least *one* black sheep in Scotland."

The third mutters, "No, technically all you see is that there is at least one sheep in Scotland that is black *on at least one side.*"

Everything we perceive and predict rests on assumptions, even the solipsistic view that challenges the existence of anything outside one's own mind. For instance, there's no guarantee that the Sun will rise tomorrow or that your shoes won't run away by themselves or that for every action there will always be an equal and opposite reaction. You choose your assumptions, bearing in mind that as with anything else, past performance is no guarantee of future results.

Like any power tool, induction is easy to misuse, and we always seem to think we're better at it than we are. We imagine that just by watching a slot machine for a while, we can read its mind. We don't bother backing up our hard drive because the odds of a crash are so low. Something wonderful isn't about to happen, just because it's never happened before. There's no way there could be a traffic jam up ahead when traffic's flowing so smoothly right here. Your quiet neighbor couldn't be the mystery killer because you've got such a great talent for diagnosing psychoses in strangers from afar.

$$123 \times 9 + 4 = 1111$$
$$1234 \times 9 + 5 = 11111$$
$$12345 \times 9 + 6 = 111111$$

$$123 \times 8 + 3 = 987$$
$$1234 \times 8 + 4 = 9876$$
$$12345 \times 8 + 5 = 98765$$

$$9 \times 9 + 7 = 88$$
$$98 \times 9 + 6 = 888$$
$$987 \times 9 + 5 = 8888$$

Parts of three cute arithmetic patterns glimpsed on Professor Rothman's whiteboard (although as seen on the show, they contain a few mistakes).[1] Without actually doing any math, try using induction to predict how each pattern will continue.

Rules always seem more acceptable when there are plausible explanations to back them up, but explanations aren't required. Years of bird-watching may have led you to conclude, through induction, that all ravens are black (the rule) without telling you *why* all ravens are black (the explanation). Howard can work out a formula that gives a good estimate of his chances of dating success (the rule) without understanding exactly *why* it works (the explanation).[2]

Here, for example, are the durations of Leonard's

last three love affairs (excluding, evidently by mutual agreement, "that girl last year at Comic-Con"):

n	Name	Duration	$^{27}/_{12^n}$ days
0	Joyce	27 days	27 days
1	Leslie	≈2 days	$2^1/_4$ days
2	Penny	3 hours	$4^1/_2$ hours

(For reasons of secrecy, Raj's sister Priya is also excluded. Chronologically, she follows Joyce and precedes the girl at Comic-Con, who precedes Leslie.)

The formula in the right-hand column matches all three data points reasonably accurately, with each duration being about one-twelfth the one above it. (Penny's is a little off, but we'll allow for travel time.) We aren't told whether it applies further back in time. Did the relationship before Joyce (presumably with the "Ph.D. in French literature" mentioned by Sheldon) last ten months, and the one before that (if there was one) ten *years*?[3]

Nevertheless, with three data points fairly well matched, we can tentatively propose this formula as a rule, and if it's a rule we'd expect it to predict Leonard's future performance.

arithmetic (or linear) progression A series of numbers, each one equaling the previous one plus some constant.
geometric (or exponential or harmonic) progression A series of numbers, each one equaling the previous one times some constant.

So when Howard says "should have ended," what he means is that if this formula applies to Stephanie as well, then the next row of the table will be:

| 3 | Stephanie | $\approx$ 20 minutes | $22^{1}/_{2}$ minutes |

Therefore, the relationship with Stephanie should have ended long ago. (And the one after Stephanie can be expected to last barely two minutes.) But it didn't. Leonard's been dating her for days already. So this formula isn't the right rule, at least in Stephanie's case. We can discard the rule and try to find a new one, or we can cling to it and create an overruling rule on

---

Zero Is the Loneliest Number

The n in the formula is an exponent indicating the number of 12s that are to be multiplied together: $12^1 = 12$, $12^2 = 12 \times 12$, $12^3 = 12 \times 12 \times 12$, and so on (see "Atto Way!").

Joyce, the first woman on the list, is assigned the number 0. Counting from $n = 0$, rather than $n = 1$, is a legitimate mathematical technique,[*] and it works here because $12^0 = 1$. If this use of zero as a counting number really troubles you, you can simply multiply the 27 in the formula by an additional 12, which will then be divided right back out again: $y = (324 \text{ days})/12^n$, where $n = 1, 2, 3, \ldots$

---

[*]   Counting from zero certainly makes more sense than putting 1 p.m. before 2 p.m. but 12 p.m. before 1 p.m. — whose idea was *that*?

top of it that says that it's the right formula ... with the exception of Stephanie.

And Priya. And the girl from Comic-Con.

Another Mini-Quiz: Answer
Both: first one and then the other.
Deductive reasoning says that *if* it is true that Penny is the only woman in the world for Leonard and *if* this date is his one chance with her and *if* he blows it, then he will never get a woman. (Sheldon leaves this conclusion unspoken.) Inductive reasoning then says that since in Sheldon's experience the failure to get a woman always turns young men into lonely, bitter old men, then *if* Leonard blows it with Penny, he too will become lonely and bitter.
Although both of these arguments are logically valid, neither one is sound: the first because at least one of its premises is almost certainly false, the second because Sheldon's experience is largely limited to TV fiction. Or perhaps hyperbole and mockery are just his way of announcing that Leonard should either stop fixating on blowing it with Penny or else cancel the stupid date altogether.
A-a-and ... we're back at "should" again.

But why does it work in the first place? Why is it a geometric progression and not an arithmetic one? Why isn't it totally random? Why does it start from 27 days? Why does each relationship last $1/12$ as long as the previous one instead of, say, $1/13$?

If we had an explanation, we might feel more

confidence in the rule, but no one has offered an explanation. And we may just be stuck with that.

Part of the power of inductive reasoning is that it narrows our focus: instead of having to wade through big messy data sets, we can concentrate on analyzing the (generally tidier) rules that generate them. We've reduced our table of numbers to a simple formula (with a glaring exception in it that you could drive a Mars rover through), which at least is something. But until someone dreams up a plausible explanation for why that formula works, all we can say is that, for whatever reason, it gives approximately correct answers for a specific subset of inputs. In other words, it's a fairly good predictor . . . of the past.

EUREKA! @ CALTECH.EDU

On Location

Almost every episode of *The Big Bang Theory* includes at least one Caltech scene, whether it's in Sheldon's office, the robotics lab, or (most frequently) the cafeteria. But in the first seven seasons of the show, the only scene actually filmed on campus was Sheldon's audience with Stephen Hawking (playing himself).[4] And that was just a single interior setup.

Other production companies haven't been so shy about visiting campus, and over the years Caltech has been featured in a number of films and television shows. The fisticuffs-and-fine-dining scene in *Beverly Hills Cop*, the title sequence to *Legally Blonde*, and the entirety of *The PHD Movie* were all filmed there, along with episodes of *Modern Family, Mission: Impossible, The X-Files, The West Wing,* and *NUMB3RS* (for which professor Gary Lorden acted as math consultant).

The stealing-the-nuclear-device sequence in the remake of *Ocean's Eleven* was *not* filmed at Caltech, though the dialogue would suggest otherwise; nor was *Real Genius*, although many of the interiors were modeled on rooms at the school.

Lesser known than all of the above, but infinitely more scientifically accurate, is the award-winning educational video series created by Caltech professor David Goodstein in the mid-1980s. *The Mechanical Universe . . . and Beyond* brings Goodstein's popular introductory physics course to life through extensive use of animation, re-enactments, and state-of-the-art computer graphics. Undergraduates assisted in writing and editing the scripts and appeared as themselves in the lecture hall scenes.

Not as universally appealing as watching two toughs wipe the dining room floor with each other in the faculty club, but much more believable.

In What Universe?
The TARDIS Apartment
Wherever the apartment building is, it must be a special building indeed, something straight out of *Doctor Who*. And the biggest mystery of all is how its insides fit into its outsides. For example, a person standing almost anywhere in the stairwell and looking up would see no stairs overhead

— only light raking down. Stranger still, the short hallway by the lobby mailboxes occupies the same physical space as does the right-hand portion of the staircase that leads up to the second floor. At the end of that hallway is a door marked STAIRWAY (having mysteriously changed from MAINTENANCE between two successive episodes in season 2). Given the apparent width of the building as seen from the street, this portal evidently leads neither to the basement stairway nor to a maintenance room but directly into the building next door.[5] Most strikingly, all the apartments on Penny's side, being located above the front entrance, evidently hang out in midair. There can't be many Cheesecake Factory waitresses who live in a TARDIS.

Pasadena Mayor Bill Bogaard weighs in on this architectural anomaly:

> Pasadena is home to some of the most brilliant people in the world. We have Caltech and JPL, the Art Center College of Design and the Pasadena Bioscience Collaborative. We have the Planetary Society, the high-tech business incubator Idealab, and Innovate Pasadena, a support network of startup executives and entrepreneurs committed to making Pasadena a center for innovation.
>
> Pasadena is also a world leader in sustainable and traditional architecture. We're well represented on the National Register of Historic Places. You can find stunning examples of the work of Greene and

Greene, Gordon Kaufmann, Myron Hunt, and Wallace Neff all over town.

With this much scientific and architectural talent in the most economically and ethnically diverse city in the state, it is not surprising to me to find an apartment building that's "bigger on the inside."

1. "The Rothman Disintegration" (Season 5, Episode 17)
2. "The Hofstadter Isotope" (Season 2, Episode 20)
3. "The Codpiece Topology" (Season 2, Episode 2), "The Irish Pub Formulation" (Season 4, Episode 6), and "The Bad Fish Paradigm" (Season 2, Episode 1), respectively.
4. "The Hawking Excitation" (Season 5, Episode 21)
5. "The Financial Permeability" (Season 2, Episode 14), "The Maternal Capacitance" (Season 2, Episode 15), and "The Mommy Observation" (Season 7, Episode 18), respectively.

TWENTY-NINE
THE SINE-TIFIC METHOD

> Leonard [preparing to push a heavy box up the stairs]: Now, we've got an inclined plane. The force required to lift is reduced by the sine of the angle of the stairs — call it thirty degrees, so about half.
> Sheldon: *Exactly* half.
> Leonard: "*Exactly*" half.
> — "The Big Bran Hypothesis" (Season I, Episode 2)

We can always count on Sheldon to focus on the words and ignore the substance, but this time is unusual in that what he misreads isn't a social cue but a mathematical one.

Three of the most common systems of units for measuring angles are degrees, radians, and gon. (One degree is a little bigger than one gon and a lot smaller than one radian.) It's easy to convert between them: you just multiply or divide (see "Hexagon with the Wind"). What's less easy is keeping track of which system you're using. If you forget to say the unit name after the number, or if you inadvertently say a different unit name, you risk disaster. When you challenge a snowboarder to "do a 720" and he accepts, will he be using the same units you are? If he completes two full rotations, that's 720 degrees; if he completes one and three-quarters rotations and lands on his face, he can always claim he assumed you meant 720 gon. As for 720 radians, or a little less than 115 full rotations, that's not often seen — even in the X Games.

There are less ambiguous methods for describing angles. They use measurements that don't involve any units at all. The sine of an angle is just a pure number, as is the slope (also called the tangent or grade). The reason the slope and the sine don't involve any units is that they're fractions: the length of one side of a triangle divided by the length of another side.

Here's the general idea: We draw two lines that cross at the angle we're interested in. Now we draw a third line crossing both the other two, perpendicular to one or the other. Any line will do; the only requirement is that the angle it makes with one or the other of the lines has to be a right angle (ninety degrees).

Where the three lines intersect, they form a triangle. Label the last-drawn side A, and measure all three sides. The sine of our angle is A's length divided by the length of the triangle's longest side. The slope of our angle is A's length divided by the length of the remaining side.

Here's another way to visualize it: Take a straight ladder and lean it up against a wall. Lean it at any angle you like, so long as its upper end rests against the wall. Now measure three things: how high above the ground the upper end of the ladder is, how far out from the wall the ladder's base is, and the length of the ladder itself. These three measurements make the three sides of a triangle, with a right angle where the ground meets the wall. The ladder is resting at some angle to the ground: zero degrees if it's lying flat on the ground, ninety degrees if it's upright against the wall, or some number in between if it's on a slant. That angle is the angle we're interested in, and we can work out its sine and slope just from the three lengths we've measured.

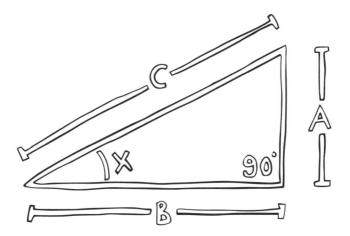

The sine of the angle **x** is **A** (the vertical difference) divided by **C** (the ladder's length). The slope of **x** is **A** (the vertical difference) divided by **B** (the horizontal difference).

The sine of our angle is the distance up the wall (from the ground to the ladder's upper end) divided by the ladder's length. It works out to a fraction between zero and one. If the ladder is four feet long and its top is resting one foot up the wall, that's a sine of 0.25 ($^1/_4 = 0.25$).

The slope of our angle is the distance up the wall divided not by the ladder's length but by the distance along the ground (from the wall to the ladder's base). It can have any value and is often expressed as a percentage. If the ladder's base is two feet from the wall and its top is resting one foot up the wall, that's a 50% slope ($^1/_2 = 50\%$). If the two distances are equal, that's a 100% slope. If the base is one foot from the wall and the top is two feet up the wall, that's a 200% slope ($^2/_1 = 200\%$).

When the ladder is standing straight up and down, with both its top and its base against the wall, the distance up the wall is the same as the ladder's length, while the distance along the ground is zero. The sine of the angle (the distance up the wall divided by the ladder's length) is one, while the slope of the angle (the distance up the wall divided by the distance along the ground) is infinity. When the ladder is lying flat, with both its top and its base on the ground, the distance along the ground is the same as the ladder's length, while the distance up the wall is zero. The sine of the angle is zero, and the slope is 0%. In between these positions (when the ladder is slanted against the wall), the distance up the wall and the distance along the ground are both less than the ladder's length but greater than zero. In those cases, the sine is some fraction between zero and one, while the slope is some number between 0% and infinity. (For downward-pointing angles, the sine can go as low as negative one and the slope can go as low as negative infinity.)

The length of the ladder doesn't matter in the slightest. If you were to use a longer ladder, the distances you measured would be correspondingly longer. Dividing these distances would give you exactly the same sine and slope as before. (See what happens in this diagram when we replace a ladder of length **C** with a ladder of length **F**.)

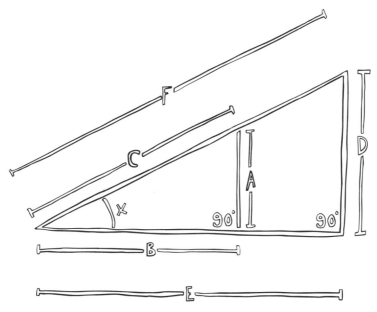

$A/c = D/F$ = the sine of the angle x. $A/B = D/E$ = the slope of the angle x.

Who cares about sines, you're wondering? Well, Sheldon and Leonard, for a start. They're pushing a box up the stairs, and the sine tells them how much altitude they'll gain after pushing it a certain distance. The 30-degree angle of their staircase has a sine of (thank you, Sheldon) *exactly* one-half, meaning that to raise the box one foot vertically, they'll have to push it two feet along the stairs. That's twice as far as they'd have to push it if they were hoisting it vertically — but of course, it'll be much easier.

And who cares about slopes? Well, the architect who designed their apartment building, for one: the slope says how much horizontal space a staircase of a given height will take up. A 30-degree angle has a slope of about 58%, meaning that the stairs take up roughly one foot of horizontal space for every seven inches they go up vertically. Judging by the size of Leonard and Sheldon's living-room window, the vertical distance from one floor to the next must be at least fifteen feet. To gain that amount of elevation, one flight of stairs would have to cover a horizontal distance of twenty-six feet or more, not including the extra space required for landings. If the stairs didn't wrap around the elevator, they'd barely fit inside the building.

Angles, sines, and slopes all scale up and down together, but at different rates. Let's say our ladder is making an angle of just a few degrees with the ground. After we measure all the distances and work out the sine and slope, we slide the base a bit closer to the wall — just enough so that the ladder makes an angle of twice as many degrees as before. Its top will creep upward as we do this. The angle has doubled, the distance along the ground has decreased, and the distance up the wall has increased. (The ladder's length, of course, remains the same.) Now when we do the division, we get a bigger sine and a bigger slope, but they're not necessarily twice as big.

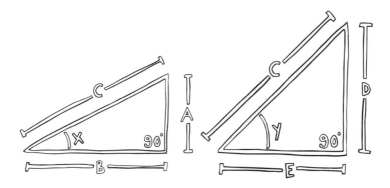

Angle *y* is bigger (in degrees) than angle *x*, and the distance **D** is bigger than **A**, while **E** is smaller than **B**. So *y*'s sine (= $^D/c$) and slope (= $^D/E$) are bigger than *x*'s sine (= $^A/c$) and slope ($^A/B$).

In the drawing above, the angle *y* is twice as big as the angle *x*, but **D** isn't quite twice as long as **A**, and **E** is shorter than **B**. So *y*'s sine (= $^D/c$) isn't quite twice *x*'s (= $^A/c$), while *y*'s slope (= $^D/E$) looks to be more than twice *x*'s (= $^A/B$).

Every angle between zero and ninety degrees has its own particular sine and its own particular slope. If you know the angle, you can look up (or compute) the sine or the slope, and vice versa. But you might not need to.

Handyman's Sine Board

Builders know some clever shortcuts for estimating angles. To find the slope of a piece of rough ground:

- Rest one end of an eight-foot piece of lumber (a common length) on the ground.
- Raise or lower the other end until the board is level.
- Measure the height of this end above the ground.

This distance (in inches) is just about the same as the slope of the ground (in percent). (The reason this trick works is that eight feet is just a few percent shy of 100 inches. You can improve the accuracy by adding an inch for every two feet measured.)

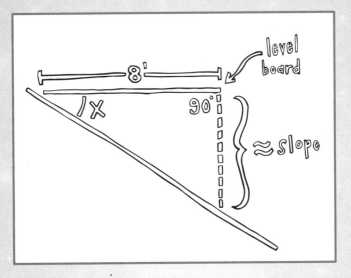

Measuring straight down: inches ≈ slope (in percent)

To find the angle of a sloping board:
- Find the point that's five feet along the board from its lower end and measure its height.
- If the height is more than a yard (36 inches), add two inches.

Up to an angle of 45 degrees, this measurement (in inches) gives a number that's within one degree of the angle. For angles greater than 45 degrees, measure sideways and subtract the result from 90. (The reason this trick works is more complicated. Don't worry too much about it.*)

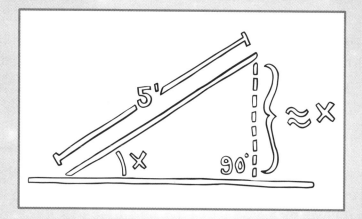

Measuring straight down: inches ≈ angle (in degrees)

---

*   All right, fine: the trick works because the sine of a small angle is a bit less than the number of radians in the angle and because the number of inches in five feet is a bit more than the number of degrees in a radian. We said don't worry about it.

Let's take a look at an analog (old-style) clock face for a minute. (Technically, this may take more than a minute — and there's no point in trying to hide that fact, since we're looking at a clock.) There are usually twelve tick marks around the circle, indicating hours. In one hour, the hour hand moves from one tick mark to the next (30 degrees) while the minute hand makes a complete 360-degree rotation, passing by all twelve of them.

The angle between adjacent hour markings, 30 degrees, is an old friend to mathematicians and engineers, though not because of anything having to do with clocks. Seeing 30 degrees in the midst of an unwieldy calculation is rather like spotting a familiar tree in the midst of a tropical rain forest.* The reason is that most angles have gnarly sines. Take 60 degrees: such a nice, round, familiar number, and yet its sine is an ugly decimal value (a little more than 0.866025). Another is 45 degrees: its sine is 0.707106 and a bit.** But 30 degrees is special. Its sine is *exactly* one-half — a clean, friendly fraction. Whatever diagonal distance you travel when climbing a 30-degree staircase, you gain exactly half that distance in altitude.

You might think it should be something smaller because 30 degrees seems like such a skinny angle. If you lay down a six-foot-long box on those stairs, will its uphill end really be three feet (= 6 feet × $^1/_2$) — *exactly* three feet — above its downhill end? It will. The sine of 30 degrees really is $^1/_2$.

---

* When Sheldon asks Penny in the first-year season closer, "Would you be open to rotating the couch clockwise thirty degrees?" he's obviously looking for any scrap of solace.[1]

** The sine of 45 degrees, usually rounded down to 0.707, is dear to the hearts of engineers, since 45-degree angles are such a common sight in our world. There's a rumor that Boeing named their first passenger jet airliner the 707 because they wanted to commemorate the fact that the wings are swept back at a 45-degree angle. However, that's an urban myth: they aren't, and they didn't.

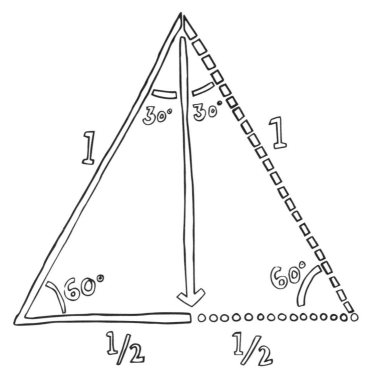

Proof that the sine of 30° = ¹/₂. All three angles of an equilateral triangle measure 60 degrees, and the dotted half-of-a-side is half as long as the dashed side.

At two o'clock, the hour hand makes an angle of 30 degrees with an imaginary horizontal line. In this diagram, the vertical dotted line connects that line to the end of the hour hand. Since the sine of 30° = ¹/₂, the dotted line is exactly half as long as the hand, even though it may appear shorter. This is true regardless of the length of the hand.

So when Leonard says that the amount of effort to slide the box up the stairs will be *about* half the effort needed to go vertically, that's because the angle is *about* 30 degrees. Sheldon, missing that second "about," takes offense at the notion that the sine of exactly 30 degrees could be anything other than exactly ¹/₂ and promptly offers his tedious correction.

Jiba Jabber

The peculiar word *sine* is a somewhat unfortunate choice because (at least in English) it sounds exactly like another mathematical term, *sign*, which usually means the + or − that tells you whether something is greater or less than zero. True, it's not the only confusing word in math and science (you might still be shaking your head over degrees, minutes, and seconds), but *sine* is an oddity for another reason, which is that it comes to us courtesy of a nine-hundred-year-old mistranslation.

In the twelfth century, a scholar known to history as Gerard of Cremona was preparing a Latin translation of an Arabic translation of an Indian book on geometry. He came to the word *jiba*, a transliteration of a Sanskrit word meaning "line slicing through a circle" or "bowstring." Perhaps understandably, Gerard mistook *jiba* for a completely different word (*jaib*), written identically in Arabic but meaning (among other things) "fold in a garment." In Latin, "fold in a garment" is *sinus*, so that's the word he used. When Gerard's translation was later adapted into English, his *sinus* was reinvented as *sine*.

With the addition of a circle, it's easy to see the connection from sine to (one-half of a) bowstring. All that's missing is a little archer.

Actually, *sinus* in Latin can mean not only "fold in a garment" but also "bay," "pocket," "hollow space within the skull bones," or "piece of land within one bend of a meandering river." You can see the connection between these concepts: they all signify a scooped-out region. But none of them looks anything like a line slicing through a circle. Oops, Gerard.

A graph that gives the sine for every angle is called a sine curve, sine wave, or sinusoid. This shape is associated with many physical processes, such as the momentum waves illustrated in "It's a Fuzzy Old World." Interestingly, a sinusoid curves back and forth in a way that very much resembles the bends in a meandering river. By some strange quirk, the graph mimics the totally unrelated thing it was accidentally named for.

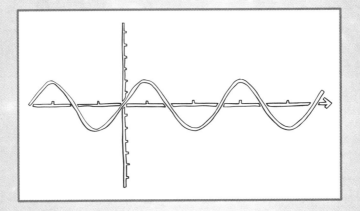

Gerard never got to see what a sine curve looks like; the first one was graphed long after he himself had disappeared around the river bend. But how nicely its little pockets and bays vindicate his own unfortunate linguistic meanderings!

How do we know that the stairs in the apartment building go up at about 30 degrees? You could measure the angle of the banister as it passes behind the elevator, but we don't have to measure it at all; by design, 30 degrees is the pitch of most staircases in the U.S. and Canada. A lower-angled staircase would take up too much horizontal space; a steeper one would make for too stiff a climb. Go ahead: find a ruler and a staircase and measure for yourself. The diagonal distance from the lip of one tread to the lip of the next will be just about twice the height of one riser.

A staircase is an example of a labor-saving device called an inclined plane. Sliding the furniture box up the stairs takes less force than hoisting it vertically, at the expense of having to move it a longer distance.

inclined plane A mechanism for redirecting an object's motion by means of a sloping surface such as a ramp, the angled side of a wedge or chisel, or the thread of a screw.[2] The inclined plane is an example of a simple machine: a basic device (such as a pulley, lever, or wheel) for redirecting or multiplying a force. (This is probably not what you think of when you hear the word *machine*.) Inclined planes work via a trade-off: the force needed goes down, but the distance traveled goes up.

Because work is defined as force times distance (the heavier something is or the higher you're lifting it, the more work you're doing), using a simple machine doesn't change the total amount of work being done, as Sheldon points out in at least two different contexts in the first season alone:

✳ While hauling a more than 200-pound replica of the H.G. Wells time machine up the stairs: "It's the same amount of work no matter how fast you go. Basic physics."[3]
✳ While lecturing Penny on how a light-year is a unit of distance, not of time: "Foot-pound

has the same problem. That's a unit of work, not of weight."[4]

But simple machines do reduce the force required, which is an important consideration if you're a couple of physicists without "a dolly, or lifting belts, or any measurable upper body strength." Would it be easier if they could somehow *drive* the box up the stairs? Maybe take it up on Howard's scooter? Probably not. Surprisingly, while the 58% slope of a 30-degree staircase is easy enough to climb on foot (in the absence of a working elevator), it would be almost too steep to drive a vehicle up safely. With a few freakish exceptions, the steepest streets in the world have slopes of much less than 40% — an angle of about 20 degrees. Trains start to run into difficulty on uphill grades of more than about 1% (half a degree).

Is all of this trigonometry strictly necessary just to get a six-foot-long furniture box up the stairs? You don't need to know anything about sines and slopes to understand that sliding it up the carpeted stairs is easier than carrying it. The only difference, and it's a pretty insignificant one, is that you wouldn't be able to calculate how much effort you'd be saving. Oh — and you wouldn't know that the slant of the stairs means that the box's upper end is exactly three feet (= 6 ft. × sine of 30°) higher than its lower end. But when there's work to be done, nobody likes a know-it-all anyhow.

We Built the Pyramids

The ancient Egyptians didn't have sines, cardboard boxes, or carpeted stairs. Yet somehow they lifted huge stone blocks and piled them into monuments, obelisks, and pyramids. A system of ramps could have been used to roll the blocks for the pyramids into position. The shallower the angle, the longer the distance needed to reach a given height but (thanks to sines) the lower the load. And ramp-and-roller technology undoubtedly existed forty centuries ago.

But the obelisks are a different matter. How do you strong-arm a solid stone column into a vertical position from the ground?

History buff Maureen Clemmons wondered whether wind power could have done the job. She mentioned the possibility to Caltech aeronautics professor Mory Gharib, who put it to the test. He built a fifteen-foot concrete obelisk, laid it down on its side, and harnessed a kite to its top.

On a moderately gusty day, the kite raised the three-ton monstrosity into a vertical position in only half a minute. If it seems implausible that a single kite could lift that much weight, keep in mind that the kite didn't need to hoist the entire obelisk; it only needed to lift *one end*. The other end remained on the ground, taking most of the obelisk's weight. And sines did the rest.

Savings Decreased More than 10% Off Lowered Price Reduction

Leonard misspeaks slightly when he says that the force "is reduced by the sine of the angle." He means "is reduced by *one minus* the sine of the angle." To see this, consider a different angle: say, 50 degrees. That would make for some ridiculously steep stairs, and sliding the box up them would be nearly as hard as hoisting it vertically. The sine of 50 degrees is about $3/4$, but that doesn't mean the force required would be reduced by $3/4$; it would be reduced by only $1/4$.

But he's stressed, he's rushed, and he's not an architect, so we can forgive him. Luckily for him, Sheldon doesn't catch it.

---

1. "The Tangerine Factor" (Season 1, Episode 17)
2. Screws are heavily discussed in this context in "The Desperation Emanation" (Season 4, Episode 5).
3. "The Nerdvana Annihilation" (Season 1, Episode 14)
4. "The Tangerine Factor" (Season 1, Episode 17)

THIRTY
FOUR AND A HALF LIVES

Sheldon: Just like Schrödinger's cat, your potential relationship . . . can be thought of as both good and bad. It is only by opening the box that you'll find out which it is.
— "The Tangerine Factor" (Season I, Episode 17)

This isn't the only time we'll hear about Schrödinger's cat from Sheldon, but his explanations never go much beyond something about some cat that's both alive and dead at the same time. Surely we can do better.

Schrödinger's cat (with a \sh\ sound) isn't a cat belonging to Schrödinger; it's a hypothetical experiment. And it's not an experiment anyone would actually carry out; there wouldn't be any point. It's a thought experiment designed to explore an apparent paradox about the nature of reality. Namely, if an event produces no detectable effect whatsoever, it's meaningless to say that it has occurred. It's also meaningless to say that it hasn't. It's like the tree that falls in the woods with no one to hear it, but considerably more mind-bending.

> **thought experiment** An experiment you only think about.

If you toss a coin many times, the laws of probability ensure that it will land on heads about half the time. If you only toss it once, you don't know which way it will land, but since a tossed coin is governed by physical laws that are well understood, you *could* (in principle) inspect the starting conditions very carefully — how you hold the coin, which side is facing up, how high you toss it, the spin speed, the windage, and so on — do a lot of math, and be able to say how it will land.

Taking measurements of certain physical phenomena — detecting the position of an electron within an atom, say, or waiting until an atomic nucleus spontaneously decays radioactively — is a similar process. As with tossing a coin, the overall distribution of outcomes over many observations follows a pattern of probability. But unlike with a coin, there's no way to predict the outcome of any individual measurement, no matter how much information we gather ahead of time; the rules governing these types of processes (if there are any) are simply hidden from us. Every time we take a measurement we'll get a result, but we can never predict what that result will be. An electron appears where it "wants" to appear. An atom decays when it "wants" to decay. If science is a search for causes (see "KISS and Tell"), each of these is an example of an effect that has no known cause.

To account for this essential unpredictability, the best we can do is to think of each particle as a superposition of one or more states (see "It's a Fuzzy Old World"). A superposition of states for a particle (or for a system of particles) isn't hard to create — it's just hard to maintain. Each time the object is observed (loosely meaning "allowed to interact with anything in such a way that all states but one temporarily become impossible to observe"), the superposition collapses to a single state: the observed one.

collapse An increase (to 100%) in the probability of observing one of the states in a superposition. This happens the moment an observation (interaction) takes place: one state is observed, and the probability of observing any other state in the superposition simultaneously drops to zero. The superposition generally remains collapsed only while the interaction is taking place. Once the object is no longer interacting with anything in its environment, it is free to return to a superposition of multiple states. Multiple observations of the same object won't necessarily collapse to the same state. Something observed to be a pencil could later be observed to be Millard Fillmore — although it's fairly unlikely.

An object that isn't interacting with anything isn't in an unknown state; it has no state. Each of the states in its superposition has some probability of being the one we'll observe at our next interaction, but none of them is "the" state of the object. It's like a game of musical chairs: as long as the music is playing, no one is any more "out" than anyone else.

Superpositions are strange creatures. They shift and change to reflect the parameters and conditions of the next observation to be made. Whenever an observation is being made, only one state will be seen. Whenever an observation is not being made, the object enters a superposition again, with every state in the superposition serving as a plausible description of what might be observed at the next interaction. This ability to answer to multiple descriptions simultaneously is bizarrely unlike anything we're used to in our daily lives. If it could be magnified to a human-sized scale, we'd find it alternately maddening, bewildering, and sublime.

Imagine you have a device for probing the state of a particle: perhaps a detector that bounces a photon off a hydrogen atom and prints out a diagram showing where its electron has revealed itself to be. Each time the device makes a measurement, the electron is forced to "choose" a state, which gets reported to you. As a result, you have information you didn't have before, and your own state changes. The electron's history is now a part of your history. The collapse of a superposition has been amplified more than a million billion trillion times — literally.

It's peculiar, but is it all that exciting? True, in the moments when the electron isn't being observed there's a superposition, but it might not matter much to you because its states are nearly identical. The electron could reveal itself to be over here (big deal), or over there (big deal), or down there (big deal). A far more impressive experiment would involve some object big enough to see, something that could be placed into two different states that look dramatically unlike each other. What if you could somehow get such an object into a superposition of those states, disallow all interaction of any kind with it so that it stayed in that superposition, and wait a good long time before finally observing it and forcing the superposition to collapse?

This is the puzzle Austrian physicist Erwin Schrödinger was mulling over in 1935, and the scenario he hit upon used something much more dramatic than electrons and diagrams. The subatomic event he picked was radioactive decay, the amplifier was a cat, and instead of printers and diagrams he used life and death.

radioactive decay A process in which the nucleus of an atom spontaneously changes (decays) into a different kind of nucleus, emitting matter and/or radiation in the process. Though any nucleus has a specific probability of undergoing decay within a given period of time, the exact moment when this happens to each particular nucleus is random and entirely unpredictable.

On average, some number of the atomic nuclei in a chunk of radioactive material can be expected to decay within a given period of time. By carefully choosing the right-sized chunk of the right material, you could

> **Warm Kitty/Cold Kitty, Can You/Can't You Purr?**
> In texts discussing this seeming paradox, it's common to see
> the inside of the box illustrated as a sort of double-exposure:
> a living cat *and* a dead cat, each of them semi-transparent
> and unaware of the other. That can be misleading. There
> aren't two cats, one still with its nine lives and one with zero.
> There isn't some "averaged-out" half-dead cat with four and
> a half lives. There's only a superposition of states, each
> with a certain probability of being observed. The two states
> with the highest probabilities each contain a single cat: one
> vertical, the other horizontal.

ensure that there was a fifty-fifty chance that sometime
during the next hour, at least one of its atoms would
decay. Schrödinger acknowledged that it was possible
to imagine some "quite ridiculous cases" (*ganz
burleske Fälle*), and he gave an example involving a
hypothetical apparatus that releases cyanide gas if
any of the atoms in a small chunk of material decays.
Together with a cat, this infernal machine is placed
into a steel chamber or box, which is then sealed in
such a way that absolutely no information about what
may be going on inside can escape to the outside
world. Until the box is reopened, we benighted souls
on the outside must remain entirely ignorant of Kitty's
plight. Her life is held in subatomic hands.[1]

After an hour, we intend to open the box and become
enlightened. If we find a dead cat, we'll know that an
atom has decayed. If we find a live (and probably very

freaked-out) cat, we'll know no atom has decayed. But until then, the question is: What's inside the box?

The answer is: There is no answer.

Since what's in the box has no effect on anything that can have an effect on us, we have no basis for declaring, sight unseen, that the cat is either alive or dead. (In fact, both of those states are superpositions of many other states. The state "dead cat," for instance, is a superposition of the states "cat died one minute ago," "cat died two minutes ago," and so forth, while "live cat" is a superposition of "napping cat," "howling cat," and many others.* Any of those is a possible description of what's inside the box.)

It's not a question of whether we could peek if we wanted or whether we could pay better attention or whether we're ignoring available evidence. (You know that the Moon is always there even when you aren't looking at it — you can see the effect it has on other things.) As long as there's no physical process that could allow us to distinguish between the states "live cat" and "dead cat" (or, equivalently, "no atom has decayed" and "one or more atoms have decayed"), the inside of the box is in a superposition of those two states, from our point of view.

For most people, this is the perplexing part. Our

---

* "Live cat" and "dead cat" are not the only possible states — just the most probable ones. There are an enormous number of other states as well, each with its own (incredibly small) probability of being observed when the box is opened. They include "no cat at all," "live cat in a gas mask it wove for itself from hairballs," and "two cats and a tone-deaf iguana discussing the global economy in Esperanto."

daily experience gives us the feeling that everything's always in *some* specific state. If you dress in the dark (as Leonard often seems to do), you may not know what colors you've put on, but the Universe surely does. Your blue shirt was in one place, your red shirt in another, and when you selected one of them to wear, you de-selected the other one. Turning on the light resolves the question for you, but it was already resolved for the Universe the moment you reached for the shirt.

In the same way, it's natural to want to insist that inside the box is a cat that definitely either *is* or *ain't* and that it's only our temporary ignorance that prevents us from knowing what the "right" answer is. But easy does it there, Pandora: as long as the box remains closed, there's simply no *is* to speak of. There ain't no live cat, there ain't no dead cat, there's only live cat/dead cat possibilities.[2] Whatever's in there, being completely disconnected from everything you're connected to, is no more "right" than "wrong," no more cat than canary. And that's why the act of opening the box doesn't just reveal to us the answer — it causes there to be one.

In his pep talk to Penny, Sheldon correctly calls what's in the metaphorical box a "potential relationship." It hasn't been observed, so it can't be said to be an actual relationship and therefore can't be said to be either a good or a bad one. However, he waters down his analogy when he says that only by opening the box can Penny "find out" what's inside it. It would be more correct for him to say that there is no relationship at all until she forces it to come into being, and to commit itself to goodness or badness, by inspecting it. Maybe he's aware that she doesn't need to hear all that. Personal

Eine kleine Katzephysik

Schrödinger's thought experiment certainly has its share of vivid elements. Unfortunately, some of his particular design choices confused an already confusing issue. For one thing, there's no fixed time at which the state change would occur: an atom could decay anytime over the course of an hour. For another, one scenario produces no remarkable change in the system, while the other imposes a distressing finality. Moreover, there's no explicit need to have a fifty-fifty chance that an atom will decay; most likely Schrödinger just saw those odds as more balanced (or more sporting).

And why, oh why, did it have to be a cat?

There are undoubtedly ways to modify the setup so that: any radioactive decay that occurs is restricted to a small window of time, the two outcomes appear visibly different from the starting state and not hideously dissimilar to each other, their probabilities are very lopsided, rerunning the experiment is simple and painless, and charismatic macrofauna are not involved.

relationships are messy, complicated, self-entangled things (especially Penny's — how refreshingly simple that probabilistic cat seems now!), and she's obviously terrified. But that's no excuse for him to mislead her on the fundamentals of quantum mechanics.

Let's suppose we've opened the cat's box and seen what's inside, and it's not a pretty sight. It's natural for us to wonder: What was inside a few minutes ago? Whatever grim tableau we see before us, it doesn't seem possible that it just sprang into being at this moment. Experience tells us (and our noses confirm) that what's in there now has been in there for some time. But that's a misinterpretation. It makes no sense to talk about "what was in there" before the unveiling caused the superposition to collapse. There simply was no "in there" in there.

What about the cat; can't a cat be an observer? Surely the cat itself is aware of what's happening to it? The problem, again, is that from our point of view there wouldn't be a cat in the box, only a potential cat: a superposition of states, each containing one cat in some degree of health. But a state containing a cat is not a cat. And Sheldon's assertion that "until the box is opened the cat can be thought of as both alive and dead" confuses the matter by focusing on the cat. It's not that it's both alive and dead; it's not that it's neither alive nor dead; it's that what we have is not *it,* a cat, but *them*, two superposed states.

One reason Schrödinger's cat is only a thought experiment is the difficulty we would have in sealing off such an apparatus from the entire rest of the Universe. There would have to be absolutely no way to distinguish between the live-cat and dead-cat states before opening the box: not by sound, not by smell, not by a shift in the center of gravity, not by the transfer of even the tiniest speck of vibration or particle of heat through the walls.

But another reason no one has tried to do the experiment is that in many ways it would be pointless. (Cat-haters may disagree.) As long as the box remained sealed, it would contain two superposed states, but you'd be able to observe neither of them. There'd be no visible evidence of the existence of that superposition. The moment you opened the box and interacted with the superposition, it would collapse and you would observe one state or the other. In either case you would see no more than what you could have seen inside an ordinary (non-isolating) box. So what would be the point?

That being said, the concept is by no means fanciful. Researchers have built several mechanisms to magnify quantum events to a visible scale, including a tiny metal bar that can be in a superposition of the states "is vibrating" and "is not vibrating." But the superposition only lasts as long as there's no interaction with the environment. The moment you peek behind the curtain by making any kind of observation, you don't see a metal bar that's somehow both vibrating and not vibrating; you see a metal bar that's either vibrating or not vibrating.

Hey, I Used to Have One Just Like That

Superpositions explain every physical process whose outcome absolutely cannot be predicted, but since to an observer they appear indistinguishable from ordinary states, it's tempting to doubt their existence. Comedian Steven Wright has a routine about indistinguishability in which he claims that everything in his apartment has been stolen and replaced with an exact replica.

If this admittedly unlikely scenario happened to you (unlikely doesn't mean impossible — just mighty far-fetched; see "KISS and Tell"), how could you prove that it had? Or that it hadn't?

A question that often arises is, "Let's say an atom does decay and the cat dies. As long as the box stays closed, the 'live cat' state is still in there as well. What must that poor cat be thinking?!" (Undoubtedly something along the lines of "Please don't open the box please don't open the box please don't open the box!")

But that's misreading the nature of superposition. Such a box doesn't contain one live-cat state plus one real dead cat. From the cat's point of view (as it were), there's only a dead cat. From our point of view, there's neither a live cat nor a dead one, only a superposition of states, each containing one cat that knows nothing about any other cat in any other state. So a more honest answer, though it smacks of Zen inscrutability, would be: "What cat?"

EUREKA! @ CALTECH.EDU

No Cats Were Harmed in the Performing of This Experiment

The environment is a superposition's worst nightmare. The moment any physical interaction occurs that couldn't have occurred in one or more states, the probability of observing any of those states goes to zero. For that reason, a superposition can only be maintained by severely restricting physical interaction, such as by supercooling individual atoms or ions and suspending them in a vacuum to isolate them from heat and vibrations. But those setups are expensive and hard to maintain, and they're fragile: stray light striking a single atom can send it reeling.

A team that includes Caltech physicist Jeff Kimble has addressed the problem by suspending a ball of pure glass in a beam of light (an "optical tweezer") at room temperature. The ball is tiny, to be sure (100 million of them could fit into a flea's egg), but it still contains millions of atoms. At that size, it's big and resilient enough to be manipulated without fear of knocking it to kingdom come, and lightweight enough for the optical tweezer to overcome gravity's pull. It's also small enough to be placed into a superposition of states easily and kept there, which persists for as long as the ball remains mechanically and thermally isolated from that meddling busybody: the environment.

OUT TO LANDS BEYOND

"Technology transfer is definitely a contact sport."
Karina Edmonds (Caltech M.S. '93, Ph.D. '98) became the U.S. Department of Energy's very first technology transfer coordinator. Her job is to help the DOE make anything discovered or developed in its seventeen national laboratories available to the private sector.

Actively sharing research and development done under government contracts with private companies helps support new enterprise and creates high-tech jobs. That cheap camera in your cell phone, for instance — the one that seems to take nothing but awkward selfies — was once an expensive government program for imaging distant galaxies.

With a bachelor's degree in mechanical engineering from the University of Rhode Island, Edmonds came to Caltech to study aeronautics, minoring in materials science. After a few years in the private sector, she moved into technology transfer at JPL (NASA's Jet Propulsion Laboratory, administered by Caltech), first as a specialist and then as director. Edmonds sees her role as connecting people and innovative technologies with market needs. In a world of industrial espionage and zealously guarded trade secrets, that's a refreshing change.

1. Erwin Schrödinger, "Die gegenwärtige Situation in der Quantenmechanik," *Die Naturwissenschaften* 23, no. 48 (1935): 807.
2. (Apologies to Dave Mason and Jim Krueger.)

THIRTY-ONE

VERY HAPPILY
EVER AFTER

Leslie Winkle: You stick electrodes in a rat's brain, give him [control of the] button, he'll push that thing until he starves to death.
Leonard: Who wouldn't?
Leslie: Well, the only difference between us and a rat is that you can't stick an electrode in our hypothalamus.
— "The Hamburger Postulate" (Season 1, Episode 5)

In the heat of what passes for a courtship ritual, Leslie is actually stating a common misconception. We can forgive her since she's previously acknowledged to Leonard that "neither of us are neuroscientists" (and evidently at least one of us "are" not a grammarian, either), and neurobiologist Amy Farrah Fowler won't show up until almost three seasons later. (Somewhat alarmingly, in Amy's first full episode she admits to having volunteered for a scientific experiment along these very lines.[1]) But let's take a moment to set the record straight.

The research Leslie is undoubtedly referring to, carried out at McGill University in 1954, most likely didn't give rats *that* much pleasure. Nor did it definitively identify anything corresponding to the widespread but fanciful notion of a "pleasure center," whether in the hypothalamus or in any other specific region of the brain. And none of the experimental rats starved to death, from addiction to pleasure or for any other reason.

For one thing, the tests lasted no longer than about three hours a day, over the course of at most four days. (Surely that's much less time than you spend on Facebook, and you haven't starved yourself to death.) True, the rats were permitted to stimulate their own brains (by pressing a lever, actually, not a button), but it's not known what they experienced internally. And although some of them returned to the lever more often than others, they also managed to find other, more productive ways to spend their time. (Rats, one — humans, zero.)

Researchers James Olds and Peter Milner limited themselves to studying the correlation between the stimulation of various brain structures and the frequency of lever pressing. In their report, published in the *Journal of Comparative and Physiological Psychology*, they were careful only to consider specific behavioral results. They didn't mention sexual response at all, and they certainly didn't presume to guess what might be going through a rat's mind, let alone what it was feeling.

The way the experimenters chose to measure the state of mind of a rat was to compare how often it pressed the lever when the electrode was operational with how often it pressed it when the power was turned off. The assumption was that if the rat pressed the lever more often when the electrode was live, it was probably experiencing something pleasurable; if less often, it was probably trying to avoid something unpleasant. (Then again, they confessed to being unable to interpret the behavior of one particular rat, who seemed to love pressing the lever regardless of whether the power was on or off. Perhaps he was just really into technology.)

Eventually the rats did all succumb, not to hunger or to overstimulation but to the surgeon's knife. In order for researchers to see where exactly the electrodes had landed, the animals had to be sacrificed and their brains sectioned. From these observations, it appeared that stimulation of the septal area, located above the hypothalamus, was associated with increased lever-pressing activity. That comes nowhere near the salacious "the hypothalamus is the seat of bliss" urban legend Leslie seems to have latched onto. The raciest statement the McGill report makes on the subject is to note that stimulation of the septal area "produces acquisition and extinction curves which compare favorably with those produced by a conventional primary reward." In other words, rats were seen to operate a "stimulate-my-septal-area" lever similarly to the way they would operate a "gimme-some-cheese" lever.

It's now understood that the septal area and the nearby nucleus accumbens are principal locations for the release of dopamine, a neurotransmitter associated with pleasure. That being so, sending alternating current through the septal area might indeed flood a rat's mind with an assortment of happy buzzes, but that's still a far cry from terminal erotic ecstasy. (As for Amy's experiment involving stimulation of a starfish's "pleasure cells," it's hard to imagine any way of verifying pleasure — let alone associating it with specific cells — in a creature that has no brain.[2])

The electrodes used in this early attempt at deep brain stimulation were fashioned from enameled silver wires attached to a transformer plugged into the wall. Cut to length and inserted through a hole in the skull, they were implanted in a careful and controlled fashion, but not with micro-precision. High-resolution real-time views into the rats' brains simply weren't available, making it impossible to predict exactly where the end of the electrode would land.

Nowadays it's possible to stimulate the brain much more precisely. The technique called optogenetics involves nurturing the growth of light-sensitive proteins inside neurons in specific regions of the brain. Tiny probes are then used to stimulate those cells, not electrically but by illuminating them via millisecond-long light pulses, triggering chemical reactions. Structures without those light-sensitive proteins remain unaffected.

Could these advances in brain-stimulating implants eventually lead to the sort of terminally addictive behavior Leslie's talking about? Are we that easily enslaved by nothing more than electrical impulses flowing through our brains' reptilian cores?

Some of the "wireheads" in science-fiction author Larry Niven's Ringworld series, for instance, become addicted to their permanent electrodes to the point of dehydration and starvation. And if the "brain in a vat" hypothesis is correct (see "Coming to Think of It"), all our desires are based not on real perceptions but on a lifetime's worth of beguiling fabrications.

Oxford neuroscientist Morten Kringelbach, a specialist in the pleasure systems of the brain, offers a glimmer of hope. He points out that there's a distinction between *wanting* something and *liking* it. There are plenty of things that we like very much but that we nevertheless spend a lot of time not doing because we simply don't want to at the moment. Life is full of delights and desires competing for our attention. Many of them we find simply irresistible, at least in the short term. But rat or human, there comes a time when you decide to give the lever a rest — just for a little while — and grab a snack instead.

EUREKA! @ CALTECH.EDU
Aw, Honey, That's Just How We Guys Are

You can provoke a rat to attack by electrically stimulating a specific region of its hypothalamus with an electrode, but if you do the same to a mouse, it just freezes or runs away.

Caltech biologist David Anderson suspected that since mice have much smaller brains than rats, excess current from the electrode might be leaking out of the aggression area and into a nearby area that controls defensive behavior. Replacing the electrode technology with optogenetics gave

---

him finer control over those regions and allowed him to induce aggression in male mice without causing them to freeze.

That same region of the hypothalamus is also active during mating, and Anderson found that lowering the level of stimulation caused the mice to switch out of attack mode and turn on the charm instead. In other words, two of the animal world's most fundamental behaviors, aggression and mating — classical opposites, yet in many ways eerily similar — are controlled by one set of neurons.

So before you start boasting, "I'm a lover, not a fighter," remember: it could just be a matter of degree.

In What Universe?
Eureka!

For our last-ditch attempt at working out the location of the apartment building, we'll try what anybody else would have thought of trying first: head out to Pasadena and look around.

In fact, we don't have to; it's been done for us. In April 2012, armed with still images of the view through various windows on the show, intrepid amateur photographer James P. Miller took to the streets of Pasadena and began hunting for the location(s) where those cityscapes could be found. First he noted the apparent relationship between the dome of Pasadena City Hall and the mountains behind it, as seen from Leonard and Sheldon's apartment. This placed him on a line passing obliquely through downtown Pasadena. By searching along that line, he was able to identify additional buildings also visible through their window.

All he had to do then was to wander around that part of town for a while, testing out buildings and balconies, until he hit upon the exact position where that view could be seen: the top of a parking structure at Green Street and Hudson Avenue, just half a mile from the Caltech campus. Turning his head to the right 30 degrees (a familiar angle; see "The Sine-tific Method"), he was delighted to discover what amounts to the view out Penny's living-room window.[3]

Miller spotted a few disconnects between what he had seen on the show and what he was seeing now. For one thing, Penny's window faces not 30 degrees to the right of Leonard and Sheldon's but 90 degrees to the left. Additionally, some buildings and structures had mysteriously undergone a change of position. And while City Hall, half a mile away, was scarcely more than a dot in the distance, it fills a substantial portion of the window of apartment 4A.

But there's a perfectly simple explanation for all of this. Clearly, the apartment building has windows that incorporate prisms, zoom lenses, and state-of-the-art digital filters. It also lurks at a phony address, masquerades as a parking structure, and has been neatly folded into the fourth dimension.

Seems plausible.

1. "The Robotic Manipulation" (Season 4, Episode 1)
2. "The Junior Professor Solution" (Season 8, Episode 2)
3. http://forum.the-big-bang-theory.com/topic/289-location-of-their-apartment-building/page-3#entry32325
   shortcut: http://DaveZobel.com/-bbjm

THIRTY-TWO
AND THAT'S HOW IT'S DONE

Understanding our world: that's the bottom-line motivator for scientific inquiry. We all do it, scientists and non-scientists alike, and not just because enlightenment is more useful than ignorance, but because it's more pleasant to guess right than to be startled.

Nobelist Richard Feynman, one of Sheldon's heroes in both physics and bongo-drumming, was fascinated with our human need to understand how the world works. "Curiosity demands that we ask questions," he says in the classic *Feynman Lectures on Physics*, and that we search for connections between "things which at first sight look different, with the hope that we may be able to *reduce* the number of *different* things . . ."[1] If every new situation struck us as absolutely novel and unlike anything we'd ever seen before, life would be unbearable. Imagine being unable to spot or trust the nuggets of familiarity when meeting a stranger, driving a friend's car, entering a coffee shop (shades of the film *Memento*). Just getting up on successive mornings would be hopelessly bewildering, even terrifying. That's why our lives are a balance of trying out new things and synthesizing our experiences.

Despite Sheldon's implication, the scientific method involves a bit more than just "perform an experiment." Experimentation without observation and analysis is just fiddling. In fact, observation and analysis suggest experiments in the first place, and observation and analysis of the results of experimentation suggest new experiments, in a cycle that may repeat many times.

Experimentation costs time and money, and nobody has an infinite supply of either. Here's an example of how induction drives the scientific method:

According to Howard, all the women in his family are very short.[2] Let's suppose this leads him to predict that every woman who will ever be born into his family will also turn out to be short.

"I bet he's right," says Penny.

"I bet he's wrong," says Amy.

(Bernadette says nothing and leaves the room.)

At the next gathering of the entire Wolowitz extended family and their friends, Howard's surrounded by women. He approaches a short woman and finds out she's she's no relation; in fact, it turns out that none of the tall women in the room are Wolowitzes. His aunt and her girlfriend arrive: the aunt is short, the girlfriend towers over everybody. So far, it's looking like you can take "all present and future Wolowitz women are short" to the bank.

Now someone mentions that four more women are on their way up, but that nothing is known about them except the following facts: Bobbie is short. Carole is tall. Theda is definitely not a Wolowitz. Alison definitely is.

How can Howard use these new arrivals either to prove or to disprove his claim? Will he have to see how tall they all are? Will he have to ask them all whether they're Wolowitzes? Will they be creeped out and immediately leave? Will they ask *him* to leave? Is he doomed to an eternity of noting the height of every woman he ever meets and asking her whether she's a Wolowitz? That seems like a lot of work, and potentially not very pleasant. For Howard, either.

The answer is just around the corner.

controlled experiment An experiment that is performed at least twice, with every condition unchanged except one. The reasoning is that any difference in outcomes will probably be due to that single parameter.

Experimentation is more than just making casual one-time observations. Repeatability and modifiability are key. If repeating an experiment exactly the same way gives the same results every time and changing the conditions changes the results, that hints at a cause-and-effect relationship. But which effects are linked to which causes? That's why the scientific method relies on controlled experiments, in which only one condition changes at a time.

Yet how can any do-over ever have only a single change? At a minimum, the air currents in the room might shift; the Earth might have moved farther along in its orbit; a superposition might collapse differently. The challenge is to identify all the differences that may be relevant and limit them — or at least measure them.

hypothesis, theory, law
These three terms are often used to mark the path from "educated guess" to "absolute truth." A hypothesis represents a reasonable guess about how the world works, repeated testing elevates it to the status of a theory, and when it's widely accepted, it becomes a law. But in some sense everything is at best an educated guess; nothing can be presented as the Absolute Truth.

hypothesis: The atomic hypothesis (the contention that stuff is made of atoms) was wildly debated not long back, but nowadays it goes virtually unchallenged.

theory: No one thinks the theory of relativity is only theoretical anymore, but some people still dispute the theory

> of evolution, and string theory remains a collection of mostly unproven (and perhaps even unprovable) conjectures.
>
> **law:** Newton's law of universal gravitation (see "KISS and Tell") doesn't apply in certain circumstances, and Moore's Law of integrated circuit density (see "Making Memories") can't possibly hold true forever, nor was it ever intended to.
>
> As for the Big Bang theory, if it were a law instead . . . well, just try to picture a David E. Kelley/Chuck Lorre co-production.

For that matter, what if our most time-tested assumptions are wrong? What if, say, the strength of gravity *has* changed very slowly over time, or *is* different in different parts of the Universe, or *will* change abruptly next Tuesday? (It almost certainly hasn't, isn't, and won't, but not even to ask would be irresponsible.) Scottish philosopher David Hume recognized that all our reasoning is based on a thoroughly arbitrary assumption that Nature's under-lying laws are universal and unchanging. In fact, the notion that there are any rules in the first place is only an assumption. That misattributed old saw that defines insanity as doing the same thing and expecting different results (which did not originate with Franklin, Einstein, Voltaire, or Dr. Phil) is at best a widely held prejudice. (It didn't originate with Hume, either.)

A good portion of scientific research involves testing assumptions, or at least acknowledging them, which is why scientific papers and lab reports are so micro-detailed and methodical. They're a way of saying, "Here's what I did, and here's what I saw when I did what I did." This is true whether the researcher is an experimenter like Leonard ("I pointed a laser at a rock and it went boom") or a theoretician like Sheldon ("I predicted that the atoms inside a rock would go boom under the influence of laser light"). A report is useless unless it provides enough information to allow others to perform the same experiment (or calculations) and either reproduce the original results (which would be reassuring) or get different results (which would be intriguing).

A complete report also says, "Here's what I think caused me to see what I saw when I did what I did." That's the researcher's opportunity to take a guess at what might be going on behind Nature's perpetual veil. But it's only a guess, and since it's based on inductive reasoning, although someone might be able to prove that it's wrong, no one can ever prove that it's right. In the most dramatic case, someone else might be able to prove it's wrong.

Sheldon's embarrassment at his own occasional errors is certainly understandable, but his arrogance doesn't serve him well, as when he makes the audacious claim that a hypothesis of his "doesn't need proving."[3] The heck it doesn't. As a professional scientist, and as a human being, he needs to accept that one of his speculations might be proven wrong (see "Ro-, Ro-, Ro- Your 'Bot"). The best he can do is to write detailed reports so that when others are making their own guesses, taking their own stabs in the dark, their dark is no darker than his was.

All the precise experimentation in the world won't

support a sloppy analysis. If every time you look in a mirror you see your reflection looking directly back at you, can you conclude from this that your reflection is always looking at you no matter where you're looking? If a flea trained to jump on command stops jumping once you've pulled all its legs off, does that prove that a flea's ears are in its legs?

Additional evidence never proves an induction; it can only disprove it or support it. For that reason, it's important to keep an open mind. Brainteasers, lateral thinking puzzles, and other people's opinions force us to re-evaluate and even discard what we hold dear, and that's as it should be. Be like a teenager. Challenge everything.

Years before he played Richard Feynman in the "nearly one-man" play *Q.E.D.*, actor Alan Alda told a college graduating class, "Your assumptions are your windows on the world. Scrub them off every once in a while or the light won't come in."[4]

And remember that scientists don't have all the answers. That would make their lives unendurably boring.

They just have great questions.

Solution to the "Short Wolowitz Women" Problem

We're supposing that Howard has predicted that every woman in his family, including those not yet born, will be short. The limited information he's collected so far supports this claim, but he'd like either to prove it (to impress Penny) or to disprove it (to impress Amy).

Sorry, Penny — but he can never prove it. How could he? He can't observe every woman who will ever be born into his family. However, he could conceivably *dis*prove it. All he has to do is find one tall Wolowitz woman. Her mere existence would show that his prediction was wrong.*

Until he finds one of these hypothetical Tall-owitzes, he'll have to keep searching; so if one exists, clearly the sooner he finds her, the better. But how?

Fortunately, he doesn't have to interview every woman at the party as well as every woman who arrives forever after. Let's consider the four women who are about to enter the room: short Bobbie, tall Carole, so-not-a-Wolowitz Theda, and Alison, who so totally is.

If Bobbie (short) is a Wolowitz, that certainly supports Howard's claim, but if she's not a Wolowitz, that neither supports nor disproves his claim, which after all only talks about Wolowitzes and says nothing about *non*-Wolowitzes. So examining the short one won't help.

What about king-sized Carole? She's tall, but if she's not a Wolowitz, that again neither supports nor disproves Howard's claim. However, if she turns out to be a Wolowitz, then the claim is disproven, and he has his answer. So he needs to find out whether the tall one is a Wolowitz.

Theda barrels in next. Though we're told she's not a Wolowitz, we aren't told her height. But it doesn't matter: she's not a Wolowitz, so no matter what her height is, it will neither support nor disprove the claim. So examining the non-Wolowitz won't help.

As for Alison Wolowitz, if she turns out to be tall, that

would disprove Howard's claim, while if she's short (as he expects), that would support it. So he needs to see whether the Wolowitz is tall.

To sum up: since the only thing that could ever change Howard's theory would be the discovery of a tall Wolowitz woman, the only women he needs to spend time investigating (as it were) are those who might be tall Wolowitzes. This means he'll want to check out Carole, who is tall (to see whether she's a Wolowitz), and Alison, who is a Wolowitz (to see whether she's tall). He can ignore short Bobbie and non-Wolowitz Theda. (No doubt they're already ignoring him.)

If either Carole *or* Alison turns out to be both tall and a Wolowitz, then his claim is disproven and he can stop his female stature investigations. But if tall Carole isn't a Wolowitz *and* if Alison Wolowitz isn't tall, then his claim isn't disproven (though of course also never proven — only supported), and he'll insist on continuing. Potentially forever.

For everyone's sake, let's hope a tall Wolowitz woman shows up soon.

---

* Then again, since Howard's the one who made the claim, he's got an interest in *not* disproving it. He might be happier just finding short Wolowitzes, since each new one seems to strengthen his claim. That conflict of interest is why programmers give their software to others to test. It's also why researchers try to replicate one another's experiments — despite the disdain Leonard's mother expresses on realizing he's doing "no original research."[5]

The Earth Began to Cool

Caltech geochemist Clair Patterson was an expert on the element lead. He argued strongly for its environmental regulation and successfully fought to have it taken out of gasoline. But he's most famous for having used it to figure out the age of the Earth.

This he did in 1956, using new techniques for detecting lead isotopes, some of which are produced by radioactive decay. Patterson analyzed the lead content of a number of meteorites and ocean-floor sediments and found that they had all cooled and solidified at the same time: just over four and a half billion years ago.

Like any conscientious scientist, he took pains in his report to defend his chief assumptions; after all, no one from four and a half billion years ago is around to tell us for certain that he got it right.* While acknowledging that the newly solidified material might not have coalesced immediately to form the Earth, he points out that it must have happened soon after, or else the resulting chemical changes (see "Fizz-ics") would have altered the isotope ratios he found. "The data can be explained by other qualifying or even contradictory assumptions," the report concedes, but it goes on, "Most of these can be excluded as improbable."[6] This is Occam's razor in action: avoiding a convoluted explanation when an elegant one will do (see "KISS and Tell").

His report also addresses the most plausible arguments against his findings. The sample preparation procedures he used might have skewed the data, he notes, but by at most a percent or two. His analysis is peppered with observations like "it is unlikely" and "it is doubtful," and twice he uses the words "it seems reasonable to believe" — which is all any scientist can claim (see "Past Performance Is No Guarantee").

Patterson's method was so analytically sound and his assumptions so well defended that in the intervening decades scientists have only been able to refine his number by a tiny amount. And that's how hot lead became cool science.

---

*   On the other hand, plenty of people are around to maintain vehemently that he got it wrong — by about four and a half billion years. But *they* don't have any proof either.

ASK AN ICON: Stephen Hawking
Stephen Hawking needs no introduction.

Q: Would you consider contributing a quote for a little book on the science of TV's *The Big Bang Theory?*
Stephen Hawking [via e-mail]: At last, a show that makes physics cool
Stephen

1. Richard P. Feynman, *The Feynman Lectures on Physics*, vol. I (Reading, MA: Addison–Wesley, 1970), 2–1. http://feynmanlectures.caltech.edu/I_02.html#Ch2-S1
    shortcut: http://DaveZobel.com/-bbfp
2. "The Panty Piñata Polarization" (Season 2, Episode 7)
3. "The Cooper-Hofstadter Polarization" (Season 1, Episode 9)
4. http://graduationwisdom.com/speeches/0020-alda1.htm
    shortcut: http://DaveZobel.com/-bbaa
5. "The Maternal Capacitance" (Season 2, Episode 15)
6. Clair Patterson, "Age of meteorites and the earth," *Geochimica et Cosmochimica Acta* 10 (1956): 230.

ACKNOWLEDGMENTS

> Sheldon: I'm being given credit that I don't deserve.
> Leonard: Oh, people get things they don't deserve all the time. Look at me with you.
> — "The Romance Resonance" (Season 7, Episode 6)

Anyone can be an author without being authoritative, and the mere act of writing doesn't make what's written right. Tremendous authorial gratitude goes to the many real authorities who have graciously agreed to appear in these pages. Thanks also to the network of kind souls who helped arrange those connections: Mitch Aiken, Deborah Castleman, Mike Cohen, Hall Daily, Mark Guidarelli, Fred Hameetman, Linda McManus, Lance Norris, John Preskill, Rusty Schweickart (the younger), and Mark Wittcoff.

Hannah Dvorak-Carbone, Leonard Finegold, Gloria Jew, Dee Wroth, and Hiller and John Zobel made helpful suggestions and contributions. Kathy

Svitil, with her inimitable microscopic thoroughness, caught and corrected huge quantities of nonsense (in a big, big bucket).

Caltech's Kristen Brown, Adam Cochran, Mory Gharib, Cassandra Horii, Phil Scanlon, and Ben Tomlin provided endless support and enthusiasm.

Lynn Gammie's delightful illustrations proved to be worth far more than the nominal kiloword a piece.

David Caron, Crissy Calhoun, and the good shepherds of ECW Press saw the project all the way through from rough concept to blockbuster motion picture series. The perpetually cheerful Jen Hale, tireless advocate for the reader, never once wavered from her insistence on holding the book to a far higher standard than the author had hoped to get away with.

As the final result bears multiple fingerprints but only one set of muddy footprints, outpourings of opprobrium should be sent not to any of these fine folks but directly to bbt@DaveZobel.com

Updated information and additional resources are available at: http://DaveZobel.com/bbt

ABOUT THE AUTHOR

Dave ⊐⊔⊟⊑⊔ is constantly humbled by how little he actually knows about science. His bachelor's degree from Caltech is seen by many as stark refutation of Woody Allen's "eighty percent of success is just showing up" thing: in four years there he failed *ei*ght courses (an average of two Fs a year, if his arithmetic is correct).* One of his most vivid memories is of a spirited discussion with the legendary Richard Feynman, although truth be told it consisted mostly of a lot of whining and pleading and led only to a renewed awareness that an F is an F.

When not confronting the mysteries of the Universe in slack-jawed bewilderment — or laboring to spread that feeling to others — Zobel designs educational

---

* Looks like arithmetic wasn't one of the eight.

397

kits for Trash for Teaching, a Los Angeles non-profit that rescues landfill-bound factory discards and repurposes them as science and art supplies for schools.

He is most proud of his appearance on the math-oriented TV drama *NUMB3RS*, in which he portrays "Small Handful of Fuzzy Background Pixels #2."

Published by ECW Press
665 Gerrard Street East, Toronto, Ontario, Canada M4M 1Y2
416-694-3348 / info@ecwpress.com

Library and Archives Canada Cataloguing in Publication

Zobel, Dave H., author
    The science of TV's The big bang theory : explanations even Penny would understand / Dave
H. Zobel.

ISBN 978-1-77041-217-0 (pbk.)
Also issued as
ISBN 978-1-77090-706-5 (pdf) and ISBN 978-1-77090-707-2 (ePub)

1. Big bang theory (Television program). 2. Science — Popular works.
I. Title.

PN1992.77.B485Z63 2015   791.45'72   C2014-907633-9
                                                   C2014-907634-7

Cover design: Michel Vrana
Cover images: explosion © kevron2002/depositphotos; atom © Zinaida Ok/Shutterstock
Interior illustrations by Lynn Gammie
Printed by Marquis   5   4   3   2   1

MIX
Paper from
responsible sources
FSC® C103567
www.fsc.org

Printed and bound in Canada

GET THE EBOOK FREE!